Lecture Notes in Mathematics 1971

Editors:
J.-M. Morel, Cachan
F. Takens, Groningen
B. Teissier, Paris

T0205976

Laure Saint-Raymond

Hydrodynamic Limits
of the Boltzmann Equation

 Springer

Laure Saint-Raymond
Département de Mathématiques et Applications
Ecole Normale Supérieure
45 rue d'Ulm
75005 Paris
France
laure.saint-raymond@ens.fr

ISBN: 978-3-540-92846-1   e-ISBN: 978-3-540-92847-8
DOI: 10.1007/978-3-540-92847-8

Lecture Notes in Mathematics ISSN print edition: 0075-8434
                              ISSN electronic edition: 1617-9692

Library of Congress Control Number: 2008943981

Mathematics Subject Classification (2000): 76-02, 76P05, 82C40, 35B40, 35A99

Cover design: SPi Publisher Services

Printed on acid-free paper

springer.com

Preface

The material published in this volume comes essentially from a course given at the Conference on "Boltzmann equation and fluidodynamic limits", held in Trieste in June 2006. The author is very grateful to Fabio Ancona and Stefano Bianchini for their invitation, and their encouragements to write these lecture notes.

The aim of this book is to present some mathematical results describing the transition from kinetic theory, and more precisely from the Boltzmann equation for perfect gases to hydrodynamics. Different fluid asymptotics will be investigated, starting always from solutions of the Boltzmann equation which are only assumed to satisfy the estimates coming from physics, namely some bounds on mass, energy and entropy. In particular the present survey does not consider convergence results requiring further regularity. However, for the sake of completeness, we will give in the first chapter some rough statements and bibliographical references for these smooth asymptotics of the Boltzmann equation, as well as for the transition from Hamiltonian systems to hydrodynamics.

Our starting point in the second chapter is some brief presentation of the Boltzmann equation, including its fundamental properties such as the formal conservations of mass, momentum and energy and the decay of entropy (for further details we refer to the book of Cercignani, Illner and Pulvirenti [31] or to the survey of Villani [106]). We then introduce the physical parameters characterizing the qualitative behaviour of the gas, and we derive formally the various hydrodynamic approximations obtained in the fast relaxation limit, i.e. when the collision process is dominating. We finally introduce the main existing mathematical frameworks dealing with the Cauchy problem for the Boltzmann equation, which can be useful for the study of hydrodynamic limits : we will particularly focus on the notion of renormalized solution defined by DiPerna and Lions [44], which will be used in all the sequel.

The third chapter is devoted to some technical results which are crucial tools for the mathematical derivation of hydrodynamic limits. Note that the general strategy to rigorously justify the formal asymptotics is to proceed by

analogy, that is to recognize the structure of the expected limiting hydrodynamic model in the corresponding scaled Boltzmann equation. These tools will
therefore not be equally used in all fluid regimes. The first point to be discussed
is the implications of the entropy inequality, which provides some bound on
the (relative) entropy, as well as some control on the entropy dissipation, and
possibly some estimates on a boundary term known as the Darrozès-Guiraud
information, depending on the scaling to be considered. The second point is
to understand how these bounds, especially that on the entropy dissipation,
allow to control the relaxation mechanism, and which consequences this implies on the distribution function. Note that, for fluctuations around a global
equilibrium, such a study goes back to Hilbert [65] and Grad [59]. The last
point to be investigated is the balance between this relaxation process due
to collisions, and the other important physical mechanism, namely the free
transport : in viscous regime the global structure of the scaled Boltzmann
equation is actually of hypoelliptic type, and one can exhibit some regulariz-
ing effect of the free transport (extending for instance the velocity averaging
lemma due to Golse, Lions, Perthame and Sentis [53]).

The incompressible Navier-Stokes limit, studied extensively in the fourth
chapter, is therefore the only hydrodynamic asymptotics of the Boltzmann
equation for which we are actually able to implement all the mathematical
tools presented in Chapter 3, and for which an optimal convergence result is
known. By "optimal", we mean here that this convergence result
  - holds globally in time;
  - does not require any assumption on the initial velocity profile;
  - does not assume any constraint on the initial thermodynamic fields;
  - takes into account boundary conditions, and describes their limiting form.

We start by recalling some basic facts about the limiting system, in particular its weak stability established by Leray [70]. We then explain the general
strategy used to establish the convergence result of the renormalized solutions to the suitably scaled Boltzmann equation (which is very similar to the
weak compactness argument of Leray), as well as the main difficulties to be
overcomed.

The moment method, introduced by Bardos, Golse and Levermore [5] requires indeed to understand how one recovers the local conservation laws in
the limit, and to determine the asymptotic behaviour of the flux terms, especially of the convection terms which are quadratic functions of the moments.
In order to do so, the moments are actually proved to be regular with respect
to the space variables x by a refined version of the velocity averaging result
due to Golse and the author [56]. Furthermore the high frequency oscillating
parts of the moments, known as acoustic waves, are filtered out by a compensated compactness argument due to Lions and Masmoudi [76]. One therefore
gets a global weak convergence result ([54] or [55]) which does not require a
precise study of the relaxation or oscillation phenomena.

In the case of a domain with boundaries, one has further to take into
account the interactions between the gas and the wall, which leads to a braking

condition if the kinetic condition is a diffuse reflection, and a slipping condition if the kinetic condition is a specular reflection.

The state of the art about the incompressible Euler limit, which is the main matter of the fifh chapter, is not so complete as for the incompresible Navier-Stokes limit. Due to the lack of regularity estimates for inviscid incompressible models, the convergence results describing the incompressible Euler asymptotics of the Boltzmann equation require additional regularity assumptions on the solution to the target equations.

Furthermore, the relative entropy method leading to these stability results controls the convergence in a very strong sense, which imposes additional conditions either on the solution to the asymptotic equations ("well-prepared initial data"), or on the solutions to the scaled Boltzmann equation (namely some additional non uniform a priori estimates giving in particular the local conservation of momentum and energy).

Under these additional a priori estimates, it is indeed possible to improve the relative entropy method, so as to take into account the acoustic waves and the Knudsen layers.

The last chapter of this survey is devoted to the compressible Euler limit, and is actually a series of remarks and open problems more than a compendium of results. The main challenge is of course to understand how the entropy dissipation concentrates on shocks and discontinuities, which should be studied in one space dimension.

Paris, France   *Laure Saint-Raymond*
November 2008

Contents

1

Introduction

1.1 The Sixth Problem of Hilbert

1.1.1 The Mathematical Treatment of the Axioms of Physics

The sixth problem asked by Hilbert in the occasion of the International Congress of Mathematicians held in Paris in 1900 is concerned with the mathematical treatment of the axioms of Physics, by analogy with the axioms of Geometry. Precisely, it states as follows :

"Quant aux principes de la Mécanique, nous possédons déjà au point de vue physique des recherches d'une haute portée; je citerai, par exemple, les écrits de MM. Mach [81], Hertz [64], Boltzmann [14] et Volkmann [107]. Il serait aussi très désirable qu'un examen approfondi des principes de la Mécanique fût alors tenté par les mathématiciens. Ainsi le Livre de M. Boltzmann sur les Principes de la Mécanique nous incite à établir et à discuter au point de vue mathématique d'une manière complète et rigoureuse les méthodes basées sur l'idée de passage à la limite, et qui de la conception atomique nous conduisent aux lois du mouvement des continua. Inversement on pourrait, au moyen de méthodes basées sur l'idée de passage à la limite, chercher à déduire les lois du mouvement des corps rigides d'un système d'axiomes reposant sur la notion d'états d'une matière remplissant tout l'espace d'une manière continue, variant d'une manière continue et que l'on devra définir paramétriquement.

Quoi qu'il en soit, c'est la question de l'équivalence des divers systèmes d'axiomes qui présentera toujours l'intérêt le plus grand quant aux principes."

The problem, suggested by Boltzmann's work on the principles of mechanics, is therefore to develop "mathematically the limiting processes [. . .] which lead from the atomistic view to the laws of motion of continua", namely to obtain a unified description of gas dynamics, including all levels of description. In other words, the challenging question is whether macroscopic concepts such as the viscosity or the nonlinearity can be understood microscopically.

L. Saint-Raymond, *Hydrodynamic Limits of the Boltzmann Equation*,
Lecture Notes in Mathematics 1971, DOI: 10.1007/978-3-540-92847-8_1,
© Springer-Verlag Berlin Heidelberg 2009

1.1.2 From Microscopic to Macroscopic Equations

Classical dynamics for systems constituted of identical particles are characterized by a Hamiltonian

$$H(x, v) = \frac{1}{2} \sum_{i=1}^{N} |v_i|^2 + \sum_{i \neq j} V(x_i - x_j)$$

with V a two-body potential.

The corresponding Liouville equation is

$$\partial_t f_N(t, x, v) + \mathbf{L} f_N(t, x, v) = 0 \tag{1.1}$$

where f_N is the density with respect to the Lebesgue measure of the system at time t, and the Liouville operator is given by

$$\mathbf{L} = \sum_{i=1}^{N} \left[\frac{\partial H}{\partial v_i} \frac{\partial}{\partial x_i} - \frac{\partial H}{\partial x_i} \frac{\partial}{\partial v_i} \right].$$

For a given configuration $w(t) = (x(t), v(t))$ the empirical density and momentum (which rigorously speaking are measures) are then defined by

$$R_w(X) = \frac{1}{N} \sum_{i=1}^{N} \delta(X - x_i)$$

$$Q_w(X) = \frac{1}{N} \sum_{i=1}^{N} v_i \delta(X - x_i)$$

Macroscopic equations such as the Euler equations or the Navier-Stokes equations (which have been historically derived through a continuum formulation of conservation of mass, momentum and energy) are then expected to be obtained as some asymptotics of the equations governing these observable quantities.

1.2 Formal Study of the Transitions

The microscopic versions of density, velocity, and energy should actually assume their macroscopic, deterministic values through the *law of large numbers*. Therefore, in order the equations describing the evolution of macroscopic quantities to be exact, certain limits have to be taken, with suitably chosen scalings of space, time, and other macroscopic parameters of the systems. So the first step in the derivation of such equations is a choice of scaling.

1.2.1 Scalings

Denote coordinates by (x, t) in the microscopic scale, and by $(\tilde{x}, \tilde{t})$ in the macroscopic scale. Let $\rho = N/L^3$ be the typical density in the microscopic unit, i.e. the number of particles per microscopic unit volume. Then, if ε is the ratio between the microscopic unit and the macroscopic unit, there are typically three choices of scalings :

- the Grad limit $\rho = \varepsilon$, $(\tilde{x}, \tilde{t}) = (\varepsilon x, \varepsilon t)$;
  (The typical number of collisions per particle is finite.)
- the Euler limit $\rho = 1$, $(\tilde{x}, \tilde{t}) = (\varepsilon x, \varepsilon t)$;
  (The typical number of collisions per particle is ε^{-1}.)
- the diffusive limit $\rho = 1$, $(\tilde{x}, \tilde{t}) = (\varepsilon x, \varepsilon^2 t)$;
  (The typical number of collisions per particle is ε^{-2}.)

The Euler and diffusive limits will be referred to as hydrodynamic limits.

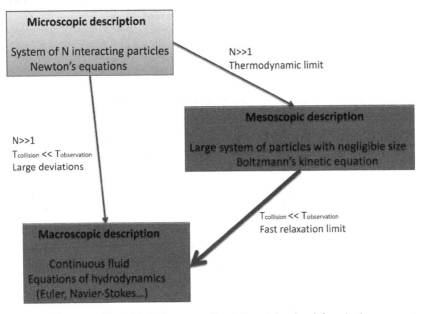

Fig. 1.1. Transitions between the different levels of description

1.2.2 Hydrodynamic Limits

To obtain hydrodynamic equations, we then differentiate the scaled empirical density and momentum and more precisely their integral agasinst any test function φ :

$$\int \varphi(\tilde{x}) R_{\omega(\tilde{t}/\varepsilon),\varepsilon}(\tilde{x}) d\tilde{x} = \frac{1}{N} \sum_{i=1}^{N} \varphi(\varepsilon x_i(\tilde{t}/\varepsilon)),$$

$$\int \varphi(\tilde{x}) Q_{\omega(\tilde{t}/\varepsilon),\varepsilon}(\tilde{x}) d\tilde{x} = \frac{1}{N} \sum_{i=1}^{N} v_i(\tilde{t}/\varepsilon) \varphi(\varepsilon x_i(\tilde{t}/\varepsilon)).$$

We get for instance

$$\frac{d}{d\tilde{t}} \frac{1}{N} \sum_{i=1}^{N} v_i(\tilde{t}/\varepsilon) \varphi(\varepsilon x_i(\tilde{t}/\varepsilon)) = -\frac{1}{N} \sum_{i=1}^{N} \varepsilon^{-1} \varphi(\varepsilon x_i) \frac{\partial H}{\partial x_i} + \frac{1}{N} \sum_{i=1}^{N} v_i \partial_i \varphi(\varepsilon x_i) \frac{\partial H}{\partial v_i}$$

$$= -\frac{1}{2N} \sum_{i=1}^{N} \nabla\varphi(\varepsilon x_i) \sum_{i \neq j} \frac{x_i - x_j}{\varepsilon} \cdot \nabla V\left(\frac{x_i - x_j}{\varepsilon}\right) + \frac{1}{N} \sum_{i=1}^{N} v_i \otimes v_i \nabla\varphi(\varepsilon x_i) + O(\varepsilon)$$

using Taylor's formula for φ, and symmetries to discard the main term.

In order to obtain the conservation of momentum in the **Euler equations** we then need to show that the microscopic current

$$-\frac{1}{2N} \sum_{i=1}^{N} \nabla\varphi(\varepsilon x_i) \sum_{i \neq j} \frac{x_i - x_j}{\varepsilon} \cdot \nabla V\left(\frac{x_i - x_j}{\varepsilon}\right)$$

converges to some macroscopic current $P = P(R, Q, E)$ depending on the macroscopic density, momentum and internal energy, in the limit $\varepsilon \to 0$. This convergence has to be understood in the sense of law of large numbers with respect to the density f_N (solution to the Liouville equation)

$$\frac{1}{N} \int f_N(t, \omega) \left| \sum_i \nabla\varphi(\varepsilon x_i) \left[\sum_{i \neq j} \frac{x_i - x_j}{\varepsilon} \cdot \nabla V\left(\frac{x_i - x_j}{\varepsilon}\right) - P(R, Q, E) \right] \right| d\omega \to 0$$

$$(1.2)$$

The key observation, due to Morrey [86], is that (1.2) holds if we replace f_N by any Gibbs measure with Hamiltonian H, or more generally if "locally" f_N is a Gibbs measure of the Hamiltonian H.

The point is therefore to establish that "locally" $f_N(t)$ is a equilibrium measure with finite specific entropy. The conclusion follows then from the *ergodicity* of the infinite system of interacting particles : the translation invariant stationary measures of the dynamics such that the entropy per microscopic unit of volume is finite are Gibbs $(\exp(-\beta H))$.

The **Navier-Stokes equations** are the next order corrections to the Euler equations. In order to derive them one needs to show that the microscopic current is well approximated up to order ε by the sum of the macroscopic current $P = P(R, Q, E)$ and a viscosity term $\varepsilon\nu\nabla Q$ (in the sense of law of large numbers).

Since there is an ε appearing in the viscosity term, proving such an asymptotics requires to understand the next order correction to Boltzmann's hypothesis. This difficulty, recognized long time ago by Dobrushin, Lebowitz and Spohn, has been overcome recently for simplified particle dynamics : the mathematical interpretation is indeed given by the fluctuation-dissipation equation which states

$$-\frac{1}{2N}\sum_{i=1}^{N}\nabla\varphi(\varepsilon x_i)\sum_{i\neq j}\frac{x_i - x_j}{2\varepsilon}\cdot\nabla V\left(\frac{x_i - x_j}{\varepsilon}\right) \tag{1.3}$$

$$= P(R_{\omega,\varepsilon}, Q_{\omega,\varepsilon}, E_{\omega,\varepsilon}) + \varepsilon\nu\nabla Q_{\omega,\varepsilon} + \varepsilon\mathbf{L}g_{\omega,\varepsilon} + o(\varepsilon)$$

for some function $g_{\omega,\varepsilon}$, where $\mathbf{L}$ is the Liouville operator. In other words, the expected asymptotics is correct only up to a quotient of the image of the Liouville operator. The image of the Liouville operator is understood as a fluctuation, negligible in the relevant scale *after time average* : for any bounded function g

$$\varepsilon\int_{0}^{t}ds f_N(s,\omega)(\varepsilon\mathbf{L}g)(\omega)d\omega = \varepsilon^2(f_N(t,\omega) - f_N(s,\omega))g(\omega)d\omega = O(\varepsilon^2)$$

and is thus negligible to the first order in ε.

In order to avoid the difficulties of the multiscale asymptotics, we may turn to the **incompressible Navier-Stokes equations** which are invariant under the incompressible scaling

$$(x, t, u, p) \mapsto (\lambda x, \lambda^2 t, \lambda^{-1}u, \lambda^2 p)$$

under which the fluctuation-dissipation equation becomes

$$-\frac{1}{2N}\sum_{i=1}^{N}\nabla\varphi(\varepsilon x_i)\sum_{i\neq j}\frac{x_i - x_j}{\varepsilon}\cdot\nabla V\left(\frac{x_i - x_j}{\varepsilon}\right) \tag{1.4}$$

$$= P(R_{\omega,\varepsilon}, Q_{\omega,\varepsilon}, E_{\omega,\varepsilon}) + \nu\nabla Q_{\omega,\varepsilon} + \mathbf{L}g_{\omega,\varepsilon} + o(\varepsilon)$$

where both the viscosity ν and the functions g are unknown. Notice that the solution to the fluctuation-dissipation equation requires inversion of the Liouville operator.

In the following two sections, we intend to describe briefly the different mathematical approaches which allow to obtain rigorous convergence results for these asymptotics. These results will be stated in a rather unformal way in order to avoid definitions and notations. We refer to the quoted publications for precise statements and proofs.

1.3 The Probabilistic Approach

The most natural approach for the mathematical understanding of hydrodynamic limits consists in using probabilistic tools such as the law of large

numbers and some large deviations principle. Nevertheless the complexity of
the problem is such that there is still no complete derivation of fluid models
starting from the full deterministic Hamiltonian dynamics.

1.3.1 The Euler Limit

Concerning the derivation of the Euler equations, what has been proved by
Olla, Varadhan and Yau [89] is the following result.

Theorem 1.3.1 *Consider a general Hamiltonian system with superstable
pairwise potential, and the corresponding stochastic dynamics obtained by
adding a noise term which exchanges the momenta of nearby particles. Suppose the Euler equation has a smooth solution in $[0, T]$. Then the empirical
density, velocity and energy converge to the solution of the Euler equations in
$[0, T]$ with probability one.*

The strength of the noise term is of course chosen to be very small so that
it disappears in the scaling limit.

The proof consists of two main ingredients. The first point is to establish
the ergodicity of the system, and more precisely the following statement : if,
under a stationary measure, the distribution of velocities conditioned to the
positions is a convex combinations of gaussians, then the stationary measure
is a convex combination of Gibbs. Noise is therefore added to the system in
order to guarantee such information on the distributions. The second point is
to prove that there is no spatial or temporal meso-scale fluctuation to prevent
the convergence (1.2).

It is based on the *relative entropy method*, so-called because the fundamental quantity to be considered is the relative entropy defined by

$$H(f|g) = \int f \log(f/g) d\omega$$

for any two probability densities f and g.

If f_N is the solution to the Liouville equation (1.1) and ψ_t is any density,
we have the following identity

$$\partial_t H(f_N(t)|\psi_t) = -\int f_N(t) \left(\psi_t^{-1}(\mathbf{L} - \partial_t)\psi_t \right) d\omega .$$

From Jensen's inequality, we then deduce that

$$\partial_t H(f_N(t)|\psi_t) \leq H(f_N(t)|\psi_t) + \log \int \psi_t \left(\psi_t^{-1}(\mathbf{L} - \partial_t)\psi_t \right) d\omega .$$

Thus, if we have

$$\frac{1}{N} \log \int \psi_t \left(\psi_t^{-1}(\mathbf{L} - \partial_t)\psi_t \right) d\omega \to 0 \tag{1.5}$$

the relative entropy can be controlled on the relevant time scale. The remaining argument can be summarized as showing that a weak version of (1.5) holds if and only if ψ_t is a local Gibbs state with density, velocity and energy chosen according to the Euler equations :

$$\partial_t R + \nabla_x \cdot (RU) = 0,$$
$$\partial_t (RU) + \nabla_x \cdot (RU \otimes U + P) = 0,$$
$$\partial_t (RE) + \nabla_x \cdot (REU - UP) = 0.$$

This is therefore a dynamical variational approach because the problem is solved by guessing a good test function.

1.3.2 The Incompressible Navier-Stokes Equations

Equation (1.4) is very difficult to solve as it requires inversion of the Liouville operator. It has been first studied by Landim and Yau [68] for the asymmetric exclusion process.

The rigorous derivation of the incompressible Navier-Stokes equations from particle systems has then been obtained in the framework of *stochastic lattice models* which are more manageable. Esposito, Marra and Yau [46] have established the convergence when the target equations have smooth solutions :

Theorem 1.3.2 *Consider a 3D lattice system of particles evolving by random walks and binary collisions, with "good" ergodic and symmetry properties. Suppose the incompressible Navier-Stokes equations have a smooth solution u in $[0, t^*]$. Then the rescaled empirical velocity densities u_ε converge to that solution u.*

Quastel and Yau [91] have then been able to remove the regularity assumption :

Theorem 1.3.3 *Consider a 3D lattice system of particles evolving by random walks and binary collisions, with "good" ergodic and symmetry properties. Let u_ε be the distributions of the empirical velocity densities. Then u_ε are precompact as a set of probability measures with respect to a suitable topology, and any weak limit is entirely supported on weak solutions of the incompressible Navier-Stokes equations satisfying the energy inequality.*

The method used to prove this last result differs from the relative entropy method, insofar as it considers more general solutions to the target equations, but - as a counterpart - gives a weaker form of convergence. One main step of the proof is to obtain the energy estimate for the incompressible Navier-Stokes equations directly from the lattice gas dynamics by implementing a renormalization group. A difficult point is to control the large fluctuation using the entropy method and logarithmic Sobolev inequalities.

It is important to note that such a derivation fails if the dimension of the physical space is less than three, meaning in particular that the 2D Navier-Stokes equations should be relevant only for 3D flows having some translation invariance.

1.4 The Analytic Approach

Here we will adopt a slightly different approach since our starting point will be the Boltzmann equation, which is the master equation of collisional kinetic theory. In other words, we will focus on the transition from the mesoscopic level of description to fluid mechanics indicated by the boldtype arrow in Figure 1.1.

Note that this will give a partial answer to Hilbert's problem insofar as Lanford [69] has proved the convergence of the hard core billiards to the Boltzmann equation in the Grad limit. Lanford's result, which is the only rigorous result on the scaling limits of many-body Hamiltonian systems with no unproven assumption, is however restrictive as it considers only short times (which will be not uniform in the hydrodynamic scalings) and perfect gases (low density limit).

For the sake of simplicity, we will consider in this section the only case when the microscopic interaction between particles is that of a hard sphere gas. We refer to the next chapter for a discussion on collision cross-sections.

1.4.1 Formal Derivations

The first mathematical studies of hydrodynamic limits of the Boltzmnn equation are due to Hilbert [65] on the one hand, and to Chapman and Enskog [33] on the other hand. Note that, in both cases, the derivations are purely formal.

Hilbert's method consists in seeking a formal solution to the scaled Boltzmann equation

$$\partial_t f + v \cdot \nabla_x f = \frac{1}{\varepsilon} Q(f, f)$$

with small variable Knudsen number ε, in the form

$$f(t, x, v, \varepsilon) = \sum_{n=0}^{\infty} \varepsilon^n f_n(t, x, v).$$

Identifying the coefficients of the different powers of ε, we then obtain systems of equations for the successive approximations f_0, $f_0 + \varepsilon f_1$, ...

Chapman-Enskog's method is a variant of the previous asymptotic expansion, in which the coefficients f_n are functions of the velocity v and of the hydrodynamic fields, namely the macroscopic density $R(t, x, \varepsilon)$, the bulk velocity $U(t, x, \varepsilon)$ and the temperature $T(t, x, \varepsilon)$ associated to f. For details, we refer to the next chapter.

Both methods allow to derive formally the Euler equations, as well as the weakly viscous Navier-Stokes equations. Let us mention however that, at higher order with respect to ε, one obtains systems of equations such as the Burnett model, the physical relevance of which is not clear. Moreover, these asymptotic expansions do not converge in general for fixed ε, and thus can represent only a very restricted class of solutions to the Boltzmann equation.

Grad [59] has proposed another, much simpler, method to derive formally hydrodynamic limits of the Boltzmann equation. This method, also called *moment method* can be actually compared to Morrey's analysis in the framework of particle systems. The first step consists in writing the local conservation laws for the hydrodynamic fields, namely the macroscopic density $R(t, x, \varepsilon)$, the bulk velocity $U(t, x, \varepsilon)$ and the temperature $T(t, x, \varepsilon)$ associated to f. The problem is then to get a closure for this system of equations, i.e. a state relation based on the hypothesis of local thermodynamic equilibrium.

1.4.2 Convergence Proofs Based on Asymptotic Expansions

Many of the early justifications of hydrodynamic limits of the Boltzmann equation are based on truncated asymptotic expansions. For instance, Caflisch [24] gave a rigorous justification of the compressible Euler limit up to the first singular time for the solution of the Euler system, which is the counterpart of the result in [89] for particle systems :

Theorem 1.4.1 *Suppose the Euler equations have a smooth solution* (R, U, T) *in* $[0, t^*]$. *Then there exists a sequence* (f_ε) *of Boltzmann solutions*

$$\partial_t f + v \cdot \nabla_x f = \frac{1}{\varepsilon} Q(f, f)$$

the moments $(R_\varepsilon, U_\varepsilon, T_\varepsilon)$ *of which tend to* (R, U, T) *as the mean free path* ε *tends to zero.*

Later Lachowicz [66] completed Caflisch's analysis by including initial layers in the asymptotic expansion, thereby dealing with more general initial data than in Caflish's original paper.

By the same method, DeMasi, Esposito and Lebowitz [42] justified the hydrodynamic limit of the Boltzmann equation leading to the incompressible Navier-Stokes equations. Like Caflisch's, their proof holds for as long as the solution of the Navier-Stokes equations is smooth, which is also reminiscent of the difficulty encountered in the framework of particle systems [46].

Theorem 1.4.2 *Suppose the incompressible Navier-Stokes equations have a smooth solution* u *in* $[0, t^*]$. *Then there exists a sequence* (f_ε) *of Boltzmann solutions*

$$\partial_t f + \frac{1}{\varepsilon} v \cdot \nabla_x f = \frac{1}{\varepsilon^2} Q(f, f)$$

which is close to the Maxwellian $\mathcal{M}_{(1, \varepsilon u, 1)}$ *with unit density and temperature, and bulk velocity* εu, *in some appropriate function space.*

Besides the solution of the Boltzmann equation so constructed that converges to a local equilibrium governed by the Navier-Stokes equation fail to be nonnegative. It could be that this problem can be solved by the same method as in Lachowicz's paper; however there is no written account of this so far.

1.4.3 Convergence Proofs Based on Spectral Results

Many other rigorous results have been obtained in a perturbative frame-work, using the spectral properties of the linearized collision operator at some Maxwellian equilibrium. Let us mention for instance the result by Nishida [88] which was the first mathematical proof of the compressible Euler limit of the Boltzmann equation. His argument used the description of the spectrum of the linearized Boltzmann equation by Ellis and Pinski [45], together with an abstract variant of the Cauchy-Kovalevski theorem due to Nirenberg and Ovsyannikov.

The more striking result based on such a spectral analysis is probably the one by Bardos and Ukai [7] concerning the incompressible Navier-Stokes limit of the Boltzmann equation. Although in the same spirit as Nishida's result, it puts less severe restrictions on the regularity of the target hydrodynamic solutions. Indeed Nishida's analysis considered analytic solutions of the compressible Euler system, and therefore was only local in time; on the contrary, the work of Bardos and Ukai considered global solutions to the Navier-Stokes equations, corresponding to initial velocity fields that are small in some appropriate Sobolev norm.

Theorem 1.4.3 *Let M be a global thermodynamic equilibrium (for instance the reduced centered Gaussian), and g_0 be some fluctuation of small norm in some appropriate weighted Sobolev space.*

Then, for any $\varepsilon \in]0, 1]$ there exists a unique global solution $f_\varepsilon = M(1+\varepsilon g_\varepsilon)$ to the scaled Boltzmann equation

$$\partial_t f_\varepsilon + \frac{1}{\varepsilon} v \cdot \nabla_x f_\varepsilon = \frac{1}{\varepsilon^2} Q(f_\varepsilon, f_\varepsilon),$$
$$f_{\varepsilon|t=0} = M(1 + \varepsilon g_0).$$

Furthermore the bulk velocity $\int M g_\varepsilon v dv$ converges uniformly to the unique strong solution of the incompressible Navier-Stokes equations.

The perturbative method employed to prove that result uses the existence of classical solutions for the incompressible Navier-Stokes equations in the Sobolev space H^l for $l > \frac{3}{2}$ with initial data small enough. The main idea by Ukai [103] is to prove that a similar theory holds for the Boltzmann equation in diffusive regime. The derivation of the Navier-Stokes limit relies then on a rigorous proof of the relation between these two theories. The point to be stressed is that exactly the same type of assumptions are made on the initial data. The Bardos-Ukai statement results then from sharp bounds on the linearized collision operator.

1.4.4 A Program of Deriving Weak Solutions

The main restrictions in the previous results are the regularity and smallness conditions on the initial data (the second assumption being possibly replaced by some restriction on the time interval on which one can prove the validity of the approximation). Such assumptions are not expected to be necessary, working with Leray solutions of the incompressible Navier-Stokes equations and with renormalized solutions to the Boltzmann equation.

That is why Bardos, Golse and Levermore [4, 5] have proposed - at the beginning of the nineties - a program of deriving weak solutions of fluid models from the DiPerna-Lions solutions of the Boltzmann equation. Their ultimate goal was to obtain a theorem of hydrodynamic limits that should need only a priori estimates coming from physics, i.e. from mass, energy and entropy bounds. In spite of significant difficulties linked to our poor understanding of renormalized solutions, this program has achieved important successes, especially in the diffusive scaling limit for which a complete convergence result is now established.

The goal of the present volume is to present an overview of these relatively recent results, and some challenging questions that remain open in that field.

The Boltzmann Equation and its Formal Hydrodynamic Limits

The kinetic theory, introduced by Boltzmann at the end of the nineteenth century, provides a description of gases at an intermediate level between the hydrodynamic description which does not allow to take into account phenomena far from thermodynamic equilibrium, and the atomistic description which is often too complex. For a detailed presentation of the various models and their derivation from the fundamental laws of physics, we refer to the book of Cercignani, Illner and Pulvirenti [31] or to the survey on the Boltzmann equation by Villani [106]. Here we will just recall some basic facts which are useful for the understanding of the problem of hydrodynamic limits.

Kinetic theory aims at describing a gas (or a plasma), that is a system constituted of a large number N of electrically neutral (or charged) particles from a *microscopic point of view*. The state of the gas is therefore modelled by a distribution function in the particle phase space, which includes both macroscopic variables, i.e. the position x in physical space, and microscopic variables, for instance the velocity v. In the case of a monatomic gas,

$$f \equiv f(t, x, v), \quad t > 0, \, x \in \Omega, v \in \mathbf{R}^3.$$

meaning that, for all infinitesimal volume $dxdv$ around the point (x, v) of the phase space, $f(t, x, v)dxdv$ represents the number of particles, which at time t, have position x and velocity v.

The function f is of course nonnegative, it is not directly observable but allows to compute all measurable macroscopic quantities which can be expressed in terms of microscopic averages, namely the local density R, the local bulk velocity U or the local temperature T

$$R(t, x) = \int f(t, x, v)dv, \quad RU(t, x) = \int f(t, x, v)vdv,$$
$$R(|U|^2 + 3T)(t, x) = \int f(t, x, v)|v|^2 dv. \tag{2.1}$$

L. Saint-Raymond, *Hydrodynamic Limits of the Boltzmann Equation*,
Lecture Notes in Mathematics 1971, DOI: 10.1007/978-3-540-92847-8_2,
© Springer-Verlag Berlin Heidelberg 2009

The distribution function f can actually be seen as the one-particle marginal of some probability density $f^{(N)}$ on the space $(\Omega \times \mathbf{R}^3)^N$ of all microscopic configurations. Of course such a *statistical description* makes sense only if the number N of particles is sufficiently large so that the gas can be considered as a continuous medium. Kinetic equations are thus obtained in the thermodynamic limit, i.e. as N tends to infinity.

From Newton's principle we can deduce a linear partial differential equation for $f^{(N)}$, the so-called Liouville equation, and then, if we neglect the interactions between particles, we obtain the following *free transport equation* for f :

$$\partial_t f + v \cdot \nabla_x f = 0, \tag{2.2}$$

meaning that particles travel at constant velocity, along straight lines, and that the density is constant along characteristic lines

$$\frac{dx}{dt} = v, \quad \frac{dv}{dt} = 0.$$

The operator $v \cdot \nabla_x$ is the classical transport operator. Its mathematical properties are much subtler than it would seem at first sight and will be discussed later. Complemented with suitable boundary conditions, equation (2.2) is the right equation for describing a classical gas of noninteracting particles. Many variants are possible. For instance, in the relativistic case, v should be replaced in (2.2) by $p/\sqrt{m^2 + (p/c)^2}$, where c is the speed of light and m is the mass of elementary particles.

Now, if the microscopic interactions between particles are described through a very long-range potential (namely in the case of electromagnetic interactions), it is enough to consider only the global effect on each particle of the interaction forces exerted by all other particles, and we get *mean field models* of the following type

$$\partial_t f + v \cdot \nabla_x f + F \cdot \nabla_v f = 0, \tag{2.3}$$

where the force F can be computed in terms of the distribution function f. For instance, in the electrostatic approximation, F is proportional to the electric field, which is itself obtained from the density $\rho = \int f dv$ by the Poisson equation.

In the case when microscopic interactions are described by some short-range potential, it is not possible to evaluate the effects of the interacting forces in a global way, using only some averaged quantities. The interactions are indeed very sensitive to the exact positions and velocities of the particles : considering for instance a system of hard spheres, i.e. of particles which collide bounce on each other like billiard balls, it is indeed easy to see that changing slightly the position of one particle may modify strongly the dynamics of the system (see Figure 2.1).

The derivation of *collisional kinetic models* requires therefore very strong assumptions to guarantee some "statistical stability" of the dynamics.

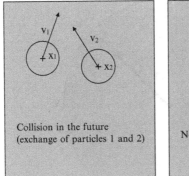 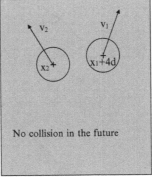

In a statistical description (which does not distinguish particles), such a perturbation has a weak effect on the binary collision.

Fig. 2.1. Instability of trajectories

2.1 Formulation and Fundamental Properties of the Boltzmann Equation

2.1.1 The Boltzmann Collision Integral

The Boltzmann equation is obtained in the thermodynamic limit $N \to \infty$ under the following conditions :

- *particles interact via binary collisions*, meaning that the gas is dilute enough that the effect of interactions involving more than two particles can be neglected. Furthermore, collisions are localized both in space and time, meaning that the typical duration and impact parameter of the interacting processes are negligible compared respectively to the typical time and space scales of the description.
  More precisely, the system has to satisfy the scaling assumption, known as *Boltzmann-Grad scaling*

$$Nd^3 << L^3, \quad Nd^2 = O(L^2),$$

  where d denotes the typical range of microscopic interactions, and L is the typical macroscopic length scale.
- *collisions are elastic*, meaning that momentum and kinetic energy are preserved in the microscopic collision process. Denoting by v', v'_* the velocities before collision, and by v, v_* the velocities after collision, the following equations have to be satisfied

$$v' + v'_* = v + v_*, \quad |v'|^2 + |v'_*|^2 = |v|^2 + |v_*|^2, \tag{2.4}$$

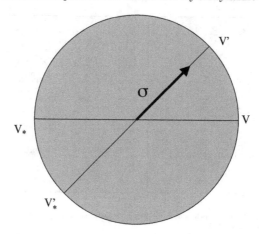

Fig. 2.2. Parametrization of elastic collisions

so that v' and v'_* can be parametrized by $\sigma \in S^2$ as shown in Figure 2.2

$$v' = \frac{v + v_*}{2} + \frac{|v - v^*|}{2}\sigma, \quad v'_* = \frac{v + v_*}{2} - \frac{|v - v^*|}{2}\sigma \qquad (2.5)$$

Note that, as the microscopic dynamics is time-reversible, the probability that (v, v^*) are changed into (v', v'_*) in a collision process is the same as the probability that (v', v'_*) are changed into (v, v_*).

- *collisions involve only uncorrelated particles*, meaning in particular that particles which have already collided are expected not to re-collide in the future. Such a chaos assumption (which implies an asymmetry between the past and the future) allows to consider that the joint distribution of velocities of particles which are about to collide is given by a tensor product (in velocity space) of f with itself.

It has been proved by Lanford in 1978 [69] that chaos is asymptotically propagated in the Boltzmann-Grad limit (at least for small times), provided that the initial probability density $f_{in}^{(N)}$ is sufficiently close to a tensor product $(f_{in})^{\otimes N}$.

The Boltzmann equation reads therefore

$$\partial_t f + v \cdot \nabla_x f = Q(f, f) \qquad (2.6)$$

where Q is a quadratic operator acting only on the v variable (first assumption), and involving tensor products (third assumption).

It is given by

$$Q(f, f) = \int_{\mathbf{R}^3} dv^* \int_{S^2} d\sigma B(v - v_*, \sigma)(f'f'_* - ff_*) \qquad (2.7)$$

where we have used the standard abbreviations

$$f' = f(v'), \quad f'_* = f(v'_*), \quad f_* = f(v_*)$$

with (v', v'_*) given by (2.5) (second assumption).

The Boltzmann collision operator can therefore be split into a *gain term* and a *loss term*

$$Q(f, f) = Q^+(f, f) - Q^-(f, f).$$

The loss term counts all collisions in which a given particle of velocity v will encounter another particle, of velocity v^*, and thus will change its velocity leading to a loss of particles of velocity v, whereas the gain term measures the number of particles of velocity v which are created due to some collision between particles of velocities v' and v'_*.

The *collisional cross-section* $B \equiv B(z, \sigma)$ is a nonnegative function depending only on $|z|$ and the scalar product $z \cdot \sigma$ (because of the microreversibility assumption), which measures in some sense the statistical repartition of post-collisional velocities given the pre-collisional velocities. It depends crucially on the nature of the microscopic interactions.

If the particles are assumed to interact via a given potential Φ, the post-collisional velocities and especially the deviation angle θ defined by

$$\cos \theta = \frac{v - v_*}{|v - v_*|} \cdot \sigma$$

can be computed in terms of the impact parameter b and relative velocity $z = v - v_*$ as the result of a classical scattering problem (see [28] for instance) :

$$\theta(b, z) = \pi - 2 \int_0^{b/s_0} \frac{du}{\sqrt{1 - u^2 - \frac{4}{|z|^2} \Phi\left(\frac{b}{u}\right)}},$$

where s_0 is the positive root of

$$1 - \frac{b^2}{s_0^2} - 4\frac{\Phi(s_0)}{|z|^2} = 0.$$

Then the cross-section B is implicitly defined by

$$B(|z|, \cos \theta) = \frac{b}{\sin \theta} \frac{db}{d\theta} |z|.$$

It can be made explicit in the case of hard spheres

$$B(|z|, \cos \theta) = a^2 |z|,$$

where a is the (scaled) radius of the spheres, and in the case of Coulomb interaction where B is given by Rutherford's formula. In the important model case of inverse-power law potentials, the cross-section cannot be computed explicitly, but one can show that

$$B(|z|, \cos\theta) = b(\cos\theta)|z|^\gamma$$

where γ depends on the power occurring in the potential, and b is a locally smooth function with a nonintegrable singularity at $\theta = 0$. The case of Maxwellian molecules corresponds to the situation when $\gamma = 0$, which is not physically relevant but enables one to do many explicit calculations in agreement with physical observations.

The nonintegrable singularity in the angular cross-section b is an effect of the huge amount of *grazing collisions*, i.e. of collisions with a very large impact parameter so that colliding particles are hardly deviated. Such a singularity appears as soon as the forces are of infinite range, no matter how fast they decay at infinity. By the way, it seems strange to allow infinite-range forces, while we assumed interactions to be localized. Anyhow, in all the sequel we shall tame the singularity for grazing collisions and replace the cross-section by a locally integrable one, which is referred to as *cut-off process*. More precisely, following Grad [59], we will assume

$$0 < B(|z|, \sigma) \le C_b(1 + |z|)^\beta \text{ a.e. on } \mathbf{R}^3 \times S^2, \text{ with } \beta \in [0, 1]$$
$$\iint_{S^2} B(z, \sigma)d\sigma \ge \frac{1}{C_b}\frac{|z|}{1 + |z|} \text{ a.e. on } \mathbf{R}^3 . \qquad (2.8)$$

In the case of a spatial domain $\Omega \subset \mathbf{R}^3$ with boundaries, the Boltzmann equation has to be supplemented with boundary conditions which model the interaction between the particles and the frontiers of the domain $\partial\Omega$. These boundary conditions have to be prescribed only on incoming trajectories, that is·on the set

$$\Sigma_- = \{(t, x, v) \in \mathbf{R}^+ \times \partial\Omega \times \mathbf{R}^3 \ / \ v \cdot n(x) < 0\} \qquad (2.9)$$

where $n(x)$ stands for the outward unit normal vector at $x \in \partial\Omega$.

The most natural boundary condition is the *specular reflection*

$$f(t, x, R_x v) = f(t, x, v), \quad R_x v = v - 2(v \cdot n(x))n(x), \quad x \in \partial\Omega. \qquad (2.10)$$

Such a condition expresses the fact that particles bounce back on the wall with a post-collisional angle equal to the pre-collisional angle. The wall is therefore considered as a perfect solid with a regular surface whose direction is precisely known. In particular, the atomistic nature of the solid and the fine details of the gas-surface interaction are not taken into account.

An alternative consists in modelling the statistical effects of the boundary irregularities, using a scattering kernel K (see [28] for further details on this topic) :

$$f(t, x, v) = \int_{v' \cdot n(x) > 0} K(v', v)f(t, x, v')dv', \quad \text{on } \Sigma_-, \qquad (2.11)$$

A particular case is the *Maxwellian reflection*

$$f(t,x,v) = \left(\int_{v' \cdot n(x) > 0} f(t,x,v')dv' \right) M_W(v), \qquad \text{on } \Sigma_-, \qquad (2.12)$$

where M_W is some fixed normalized gaussian distribution depending on the temperature of the wall. In this model, particles are absorbed and then re-emitted according to the distribution M_W, corresponding to a thermodynamic equilibrium between particles and the wall.

Of course one can combine the above conditions, which leads to more realistic models.

It is important to note that the set of characteristics relying on the singular set

$$\Sigma_0 = \{(t,x,v) \in \mathbf{R}^+ \times \partial\Omega \times \mathbf{R}^3 \ / \ v \cdot n(x) = 0\}$$

is of zero Lebesgue measure, so that it is not necessary to define the distribution function on it. (We refer for instance to the results - based on Sard's Theorem - established by Bardos in [3].)

2.1.2 Local Conservation Laws

The pre-postcollisional change of variable

$$(v',v'_*,\sigma) \mapsto (v,v_*,\sigma)$$

is involutive (since the collisions are assumed to be elastic) and has therefore unit Jacobian. Furthermore, as a consequence of microreversibility, it leaves the cross-section invariant.

Then, if φ is an arbitray continuous function of the velocity v

$$\int_{\mathbf{R}^3} Q(f,f)\varphi(v)dv$$
$$= \int_{\mathbf{R}^3 \times \mathbf{R}^3} dvdv_* \int_{S^2} d\sigma B(v - v_*, \sigma)(f'f'_* - ff_*)\varphi$$
$$= \frac{1}{2} \int_{\mathbf{R}^3 \times \mathbf{R}^3} dvdv_* \int_{S^2} d\sigma B(v - v_*, \sigma)(f'f'_* - ff_*)(\varphi + \varphi_*)$$
$$= \frac{1}{4} \int_{\mathbf{R}^3 \times \mathbf{R}^3} dvdv_* \int_{S^2} d\sigma B(v - v_*, \sigma)(f'f'_* - ff_*)(\varphi + \varphi_* - \varphi' - \varphi'_*)$$

provided that f satisfies convenient integrability conditions.

As an immediate consequence, whenever φ satisfies the functional equation

$$\varphi(v) + \varphi(v_*) = \varphi(v') + \varphi(v'_*) \qquad \forall (v,v_*,\sigma) \in \mathbf{R}^3 \times \mathbf{R}^3 \times S^2 \qquad (2.13)$$

then, at least formally

$$\int_{\mathbf{R}^3} Q(f,f)\varphi(v)dv = 0.$$

An important result in the theory of the Boltzmann equation asserts that all measurable a.e. finite functions satisfying (2.13) are linear combinations of the *collision invariants*

$$1, v_1, v_2, v_3, |v|^2.$$

The proof of this result is far from obvious; see for instance [28].

This leads to the formal *conservation laws* for the Boltzmann equation.

Proposition 2.1.1 *Let $f \equiv f(t, x, v)$ be a solution of the Boltzmann equation (2.6) that is locally integrable and rapidly decaying in v for each (t, x). Then the following local conservation laws hold :*

$$\partial_t \int_{\mathbf{R}^3} f dv + \nabla_x \cdot \int_{\mathbf{R}^3} v f dv = 0,$$

$$\partial_t \int_{\mathbf{R}^3} v f dv + \nabla_x \cdot \int_{\mathbf{R}^3} v \otimes v f dv = 0, \qquad (2.14)$$

$$\partial_t \int_{\mathbf{R}^3} \frac{1}{2} |v|^2 f dv + \nabla_x \cdot \int_{\mathbf{R}^3} \frac{1}{2} |v|^2 v f dv = 0,$$

respectively the local conservation of mass, momentum and energy.

Yet, to this date, no mathematical theory has been able to justify these simple rules at a sufficient level of generality. Even the corresponding global conservation laws in the absence of boundaries are not established. The problem is of course that too little is known about how well behaved are the solutions to the Boltzmann equation.

With the notations of the introduction for the thermodynamic fields, namely the local density R, the local bulk velocity U and the local temperature T

$$R(t, x) = \int f(t, x, v) dv, \quad RU(t, x) = \int f(t, x, v) v dv,$$

$$R(|U|^2 + 3T)(t, x) = \int f(t, x, v) |v|^2 dv,$$

and the following definition of the pressure tensor

$$P(t, x) = \int (v - U)^{\otimes 2} f(t, x, v) dv$$

these continuity equations are

$$\partial_t R + \nabla_x \cdot (RU) = 0,$$
$$\partial_t (RU) + \nabla_x \cdot (RU \otimes U + P) = 0,$$
$$\partial_t (R|U|^2 + \mathrm{tr}(P)) + \nabla_x \cdot (U(R|U|^2 + \mathrm{tr}(P)) + 2P \cdot U)$$
$$= -\nabla_x \cdot \left(\int (v - U)|v - U|^2 f dv \right),$$

where $\mathrm{tr}(P)$ denotes the trace of the pressure tensor. Note that these equations are very similar to the Euler equations for compressible perfect gases.

2.1.3 Boltzmann's H Theorem

The other very important feature of the Boltzmann equation comes also from the symmetries of the collision operator. Without caring about integrability issues, we plug $\varphi = \log f$ into the symmetrized integral obtained in the previous paragraph, and use the properties of the logarithm, to find

$$
D(f) \stackrel{\text{def}}{=} - \int Q(f,f) \log f \, dv
$$

$$
= \frac{1}{4} \int_{\mathbf{R}^3 \times \mathbf{R}^3 \times S^2} dv \, dv_* \, d\sigma \, B(v - v_*, \sigma)(f'f'_* - ff_*) \log \frac{f'f'_*}{ff_*} \geq 0
\tag{2.15}
$$

The so-defined entropy dissipation is therefore a nonnegative functional, and it can be proved that its minimizers (in the class of locally integrable functions rapidly decaying and such that $\log f$ has at most polynomial growth as $|v| \to \infty$) are Maxwellian densities, i.e. distribution functions of the following form

$$
\mathcal{M}_{R,U,T}(v) = \frac{R}{(2\pi T)^{3/2}} \exp\left(-\frac{|v - U|^2}{2T}\right)
\tag{2.16}
$$

for some $R, T > 0$ and $U \in \mathbf{R}^3$. This result is an easy consequence of the characterization of the collision invariants provided that f is continuous. In the general case, it can be proved by a nice argument due to Perthame (see [16] for instance) using the Fourier transform of the functional equation on f.

This leads to *Boltzmann's H theorem*, also known as second principle of thermodynamics, stating that the entropy is (at least formally) a Lyapunov functional for the Boltzmann equation :

Proposition 2.1.2 *Let $f \equiv f(t, x, v)$ be a solution of the Boltzmann equation (2.6) that is locally integrable and such that f is rapidly decaying in v and $\log f$ has at most polynomial growth as $|v| \to \infty$ for each (t, x). Then the following local entropy inequality holds :*

$$
\partial_t \int f \log f \, dv + \nabla_x \cdot \int v f \log f \, dv = -D(f) \leq 0.
\tag{2.17}
$$

Again this differential inequality is formally reminiscent of the Lax-Friedrichs criterion that selects admissible solutions of the compressible Euler equations. In particular, it demonstrates that the Boltzmann model has some irreversibility built in. However a considerable difference with the theory of hyperbolic system of conservations laws is that Boltzmann's H theorem provides an expression for the entropy dissipation rate in terms of the distribution function, which is local in (t, x).

2.2 Orders of Magnitude and Qualitative Behaviour of the Boltzmann Equation

The aim of this section is to give an overview of the dynamics associated with the Boltzmann equation, depending on the relative sizes of the various physical parameters. Roughly speaking, the convection phenomena are governed by the transport operator, whereas the diffusion phenomena are ruled by the collision operator. The main features of the macroscopic flow should then depend on the balance between these two terms, and especially of the ratio between the various typical length (or time) scales arising in the system.

2.2.1 Nondimensional Form of the Boltzmann Equation

Choose some observation (macroscopic) length scale l_o and time scale t_o, and a reference temperature T_o. This defines two velocity scales :

- one is the speed at which some macroscopic portion of the gas is transported over a distance l_o in time t_o, i.e.

$$\frac{l_o}{t_o} \, ;$$

- the other one is the *thermal speed* of the molecules with energy $\frac{3}{2}kT_o$, where k is the Boltzmann constant; in fact, it is more natural to define this velocity scale as

$$c = \sqrt{\frac{5}{3}\frac{kT_o}{m}}$$

m being the molecular mass, which is the *speed of sound* in a monatomic gas at the temperature T_o.

Define next the dimensionless variables involved in the Boltzmann equation, i.e. the dimensionless time, space and velocity variables as

$$\tilde{t} = \frac{t}{t_o}, \quad \tilde{x} = \frac{x}{l_o}, \text{ and } \tilde{v} = \frac{v}{c}.$$

Define also the dimensionless number density

$$\tilde{f}(\tilde{t}, \tilde{x}, \tilde{v}) = \frac{l_o^3 c^3}{N} f(t, x, v) \overset{\text{def}}{=} \frac{c^3}{R_o} f(t, x, v),$$

where N is the total number of gas molecules in a volume l_o^3, meaning that R_o is the average macroscopic density.

Finally, since the Boltzmann kernel B has units of the reciprocal product of density by time, it determines a timescale τ by

$$\int \mathcal{M}_{(R_o,0,T_o)}(v)\mathcal{M}_{(R_o,0,T_o)}(v_*)B(v - v_*, \sigma)d\sigma dv_* dv = \frac{N}{l_o^3 \tau} \, .$$

The finiteness of the above integral is ensured by Grad's cutoff assumption (2.8) on B, so that $0 < \tau < +\infty$. This is the scale of the average time that particles in the equilibrium density $\mathcal{M}_{(R_0,0,T_0)}$ spend traveling freely between two collisions, the so-called *mean free time*. It is related to the length scale of the *mean free path* λ

$$\lambda = c\tau.$$

Define the dimensionless Boltzmann kernel $\tilde{B}$ by the relation

$$\tilde{B}(\tilde{v} - \tilde{v}_*, \sigma) = R_0 \tau B(v - v_*, \sigma)$$

and set the corresponding dimensionless collision operator to be

$$\tilde{Q}(\tilde{f}, \tilde{f}) = \iint dv_* d\sigma \, \tilde{B}(\tilde{v} - \tilde{v}_*, \sigma)(\tilde{f}'_* \tilde{f}' - \tilde{f}_* \tilde{f}).$$

Then, the Boltzmann equation

$$\partial_t f + v \cdot \nabla_x f = Q(f, f),$$

can be reformulated in terms of dimensionless variables

$$\frac{l_o}{ct_o} \partial_{\tilde{t}} \tilde{f} + \tilde{v} \cdot \nabla_{\tilde{x}} \tilde{f} = \frac{l_o}{\lambda} \tilde{Q}(\tilde{f}, \tilde{f}).$$

The factor multiplying the collision integral is the inverse *Knudsen number*

$$\text{Kn} = \frac{\lambda}{l_o},$$

while the factor multiplying the time derivative is the kinetic *Strouhal number*

$$\text{St} = \frac{l_o}{ct_o}$$

(by analogy with the notion of Strouhal number used in the dynamics of vortices). Hence the dimensionless form of the Boltzmann equation is (dropping all tildas)

$$\text{St}\partial_t f + v \cdot \nabla_x f = \frac{1}{\text{Kn}} Q(f, f). \tag{2.18}$$

Before discussing the qualitative behaviour of the solution to the Boltzmann equation in terms of the relative sizes of the parameters Kn and St, let us comment a little bit on the choice of the reference scales, and introduce another dimensionless parameter which allows to compensate the arbitrariness of this choice.

A rather natural thing to do is to choose the length, time and temperature scales l_o, t_o, T_o in a way that is consistent with the geometry of the domain where the gas motion takes place, the time necessary to observe significant gas motion, and the distribution function at the initial instant of time. In this case, the ratio l_o/t_o corresponds to the *bulk velocity* u_o of the flow and the Strouhal number is nothing else than the *Mach number*

$$\text{Ma} = \frac{u_o}{c}.$$

However, considering small fluctuations around some reference flow, it may happen that the bulk velocity u_o to be studied is very small compared to the ratio l_o/t_o (which leads to some "linearized" hydrodynamics), so it makes sense to consider situations such that

$$\text{Ma} << \text{St.}$$

2.2.2 Hydrodynamic Regimes

All hydrodynamic limits of the Boltzmann equation correspond to situations where the Knudsen number Kn satisfies

$$\text{Kn} << 1.$$

Indeed, in view of Boltzmann's H theorem, one expects the distribution function to resemble more and more a local Maxwellian when $\text{Kn} \rightarrow 0$. In other words, the collision mechanism holds on a time scale which is very small compared to the observation time scale, so that one can consider that local thermodynamic equilibrium is reached almost instantaneously. This means that the Knudsen number Kn governs the transition from kinetic theory to hydrodynamics.

But there is no universal prescription for the Strouhal number in this context; as we shall see below, various hydrodynamic regimes can be derived from the Boltzmann equation by appropriately tuning the Strouhal number St.

The Compressible Euler Limit

is the easiest of all hydrodynamic limits of the Boltzmann equation at the formal level, as can be expected from the previously mentioned analogy between the system of conservation laws (2.14) associated with the Boltzmann equation, and the compressible Euler system. Indeed, as $\text{Kn} \rightarrow 0$, solutions of the Boltzmann equation behave as local Maxwellians, namely

$$f(t,x,v) \sim \frac{R(t,x)}{(2\pi T(t,x))^{3/2}} \exp\left(-\frac{|v - U(t,x)|^2}{2T(t,x)}\right)$$

for some $R(t,x), T(t,x) > 0$ and $U(t,x) \in \mathbf{R}^3$.

Therefore, passing to the limit in the local conservation laws (2.14), we get

$$
\begin{aligned}
&\mathrm{St}\partial_t R + \nabla_x \cdot (RU) = 0, \\
&\mathrm{St}\partial_t(RU) + \nabla_x \cdot (RU \otimes U + RT Id) = 0, \\
&\mathrm{St}\partial_t(R|U|^2 + 3RT) + \nabla_x \cdot \left(U(R|U|^2 + 5RT)\right) = 0,
\end{aligned}
\tag{2.19}
$$

which are the equations of hydrodynamics for perfect gases, satisfying in particular the *state relation*

$$
P = RT Id.
$$

That there is no excluded volume in this state relation is strongly linked with the Boltzmann-Grad scaling assumption $Nd^3 \ll l_o^3$, which expresses the fact that the volume occupied by the particles is negligible compared with the volume of the domain.

Furthermore, taking limits in the local entropy inequality (2.17), we obtain

$$
\mathrm{St}\partial_t \left(R \log \frac{R}{T^{3/2}} \right) + \nabla_x \cdot \left(RU \log \frac{R}{T^{3/2}} \right) \leq 0,
\tag{2.20}
$$

which is exactly the Lax admissibility condition, characterizing among the solutions of (2.19) those which are physically relevant, i.e. which satisfied the second principle of thermodynamics.

In other words, we expect the moments of the solution f to the Boltzmann equation to be approximated at order $O(\mathrm{Kn})$ by the solution to the compressible Euler equations.

A natural question is then to determine higher order hydrodynamic corrections to the compressible Euler system.

Higher Order Hydrodynamic Approximations

can be obtained by using asymptotic expansions of the distribution function in terms of the Knudsen number Kn, or in other words by seeking solutions of the scaled Boltzmann equation (2.18) as formal power series in Kn

$$
f(t, x, v) = \sum_{k \geq 0} (\mathrm{Kn})^k f_k(t, x, v),
$$

with coefficients f_k that are smooth in (t, x, v) and rapidly decaying as $|v| \to \infty$. Of course the leading order approximation f_0 is expected to be the limiting hydrodynamic distribution function, that is the local Maxwellian with thermodynamic fields satisfying the compressible Euler equations (2.19), while the successive corrections f_k account for finite Knudsen effects. Note that, depending on the exact form of the Ansatz, this process will lead to different hierarchies of PDEs.

Hilbert's expansion

$$f(t, x, v) = f_0(t, x, v) \left(1 + \sum_{k \geq 1} (\mathrm{Kn})^k g_k(t, x, v) \right)$$

is historically the older and goes back to Hilbert's fundamental paper [65] on the kinetic theory of gases. Plugging this Ansatz in the scaled Boltzmann equation (2.18), and balancing the resulting coefficients of the successive powers of Kn, one gets, as compatibility conditions to solve the hierarchy, that at each order $k \geq 1$, the hydrodynamic part of g_k satisfies the linearized compressible Euler equations (with source terms depending on g_{k-j}, for $j = 1, ..., n-1$). It seems then natural to collect all the contributions to the local thermodynamic equilibrium at leading order.

Such a variant of Hilbert's expansion was found independently by Chapman and Enskog, and is known today as *Chapman-Enskog's expansion* [33]

$$f(t, x, v) = \mathcal{M}_f(t, x, v) \left(1 + \sum_{k \geq 1} (\mathrm{Kn})^k \tilde{g}_k(t, x, v) \right)$$

where $\mathcal{M}_f$ is the local Maxwellian with same moments as f

$$\mathcal{M}_f(t, x, v) = \frac{R(t, x)}{(2\pi T(t, x))^{3/2}} \exp\left(-\frac{|v - U(t, x)|^2}{2T(t, x)} \right),$$

$$R(t, x) = \int f(t, x, v) dv, \quad RU(t, x) = \int v f(t, x, v) dv, \tag{2.21}$$

$$R(|U|^2 + 3T)(t, x) = \int |v|^2 f(t, x, v) dv,$$

and the fluctuations $\tilde{g}_k$ are functions of v depending on (t, x) through $R(t, x)$, $U(t, x)$ and $T(t, x)$ and their partial x-derivatives evaluated at (t, x). Note that, at variance with Hilbert's expansion, Chapman-Enskog's Ansatz requires knowing in advance that the successive corrections to the compressible Euler system (2.19) within any order in Kn are systems of local conservation laws.

The first correction to the compressible Euler equations is then given by

$$\mathrm{St} \partial_t \mathcal{M}_f + v \cdot \nabla_x \mathcal{M}_f = -\mathcal{M}_f \mathcal{L}_{\mathcal{M}_f}(\tilde{g}_1),$$

or equivalently

$$\mathrm{St} \partial_t \left(\log R - \frac{3}{2} \log T - \frac{1}{2T} |v - U|^2 \right) + \nabla_x \left(\log R - \frac{3}{2} \log T - \frac{1}{2T} |v - U|^2 \right) \cdot v$$
$$= -\mathcal{L}_{\mathcal{M}_f}(\tilde{g}_1),$$

where $\mathcal{L}_{\mathcal{M}_f}$ denotes the linearization of the collision operator at the local Maxwellian $\mathcal{M}_f$. Then, using the properties of the linearized collision operator $\mathcal{L}_{\mathcal{M}_f}$ (to be studied in the next chapter), namely the fact that it is a

Fredholm operator, one obtains the compressible Navier-Stokes system with $O(\text{Kn})$ dissipation terms :

$$\text{St}\partial_t R + \nabla_x \cdot (RU) = 0,$$
$$\text{St}\partial_t(RU) + \nabla_x \cdot RU \otimes U + RT Id) = \text{Kn}\nabla_x \cdot (\mu(R,T)DU) + O(\text{Kn}^2),$$
$$\text{St}\partial_t(R|U|^2 + 3RT) + \nabla_x \cdot \big(U(R|U|^2 + 5RT)\big) = \text{Kn}\nabla_x \cdot (\kappa(R,T)\nabla_x T)$$
$$+\text{Kn}\nabla_x \cdot (\mu(R,T)DU \cdot U) + O(\text{Kn}^2),$$

(2.22)

where DU denotes the traceless part of the deformation tensor

$$DU = \frac{1}{2}(\nabla_x U + (\nabla_x U)^T) - \frac{1}{3}(\nabla_x \cdot U)\,Id,$$

and the diffusive coefficients, namely the viscosity $\mu \equiv \mu(R,T)$ and the heat conductivity $\kappa \equiv \kappa(R,T)$, are defined in terms of the linearized collision operator $\mathcal{L}_{M_f}$.

We then deduce formally that the solution to the Navier-Stokes equations is close to the moments of the solution f to the Boltzmann equation at order $O(\text{Kn}^2)$.

Such a process can be iterated in order to get further corrections to the Navier-Stokes system, which leads to a hierarchy of hydrodynamic models (note however that their well-posedness requires a convenient truncation algorithm, as that proposed recently by Bobylev and Levermore [13]).

The Main Qualitative Features of the Hydrodynamic Flows

governed by the Boltzman equation can therefore be characterized in terms of the nondimensional parameters introduced at the beginning of this section, namely the Knudsen, Strouhal and Mach numbers Kn, St and Ma.

The previous results are summarized in Figure 2.3.

2.2.3 Corrections to Hydrodynamic Approximations

Furthermore we are also able to estimate by how much the solutions to the scaled Boltzmann equation deviate from their hydrodynamic approximations, at least inside the domain Ω (see Figure 2.4).

The *adiabaticity* of the gas is indeed measured in terms of the Knudsen number Kn. In a gas close to local thermodynamic equilibrium, the deviation from the hydrodynamic approximation is given by an entropic relaxation on a time scale of order Kn.

The *compressibility* of the fluid is then measured in terms of the Mach number Ma. In a weakly compressible fluid, the deviation from the incompressible approximation is given by compression/decompression waves, also called acoustic waves, oscillating on a period of order Ma.

The *viscosity* of a perfect gas is measured in terms of the Reynolds number $\text{Re} = \text{Ma}/\text{Kn}$. In a weakly viscous fluid, the deviation from the hyperbolic approximation is given by a small diffusion which smoothes the shock profiles on length scales of order $1/\sqrt{\text{Re}}$.

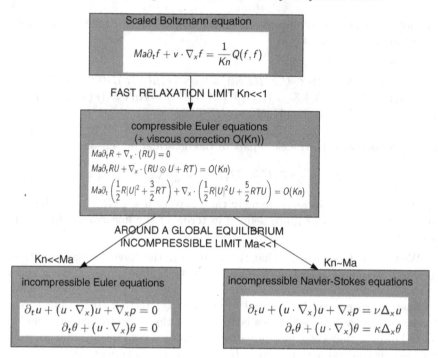

Fig. 2.3. Hydrodynamic models for rarefied gases

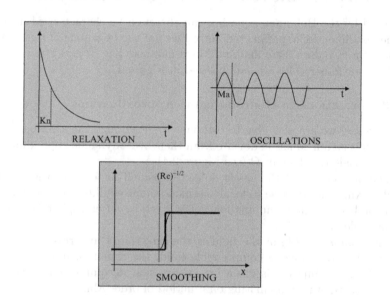

Fig. 2.4. Corrections to hydrodynamic approximations

2.2.4 Taking into Account the Boundary

It remains then to understand what happens in the vicinity of the boundaries $\partial\Omega$, which can be either exterior boundaries or obstacles.

Let us first recall that, at the microscopic level, the interaction between the gas and the boundaries is modelized phenomenologically by Maxwell's condition. If the boundary is perfectly smooth, the reflection is specular. If the boundary is rough, one further introduces some diffusion by a scattering operator, which is a relevant approximation when considering large length scales compared to the boundary irregularities. The roughness of the boundary is then measured by a supplementary non-dimensional parameter $\alpha \in [0,1]$, called the *accommodation coefficient*. More precisely the balance between the outgoing and incoming part of the trace of f states

$$f_{|\Sigma_-} = (1-\alpha)Lf_{|\Sigma_+} + \alpha K(f_{|\Sigma_+}) \quad \text{on } \Sigma_- \tag{2.23}$$

where we recall that the outgoing/incoming sets Σ_+ and Σ_- at the boundary $\partial\Omega$ are defined by

$$\Sigma_{\pm} = \{(x,v) \in \partial\Omega \times \mathbf{R}^3, \quad \pm n(x) \cdot v > 0\} \tag{2.24}$$

denoting by n the outward normal on $\partial\Omega$.

The local reflection operator L is given by

$$Lf(x,v) = f(x, R_x v) \tag{2.25}$$

where $R_x v = v - 2(v \cdot n(x))n(x)$ is the velocity before the collision with the wall. The diffuse reflection operator K is given by

$$Kf(x,v) = M_W(v) \int_{v' \cdot n(x) > 0} f(x,v') (v' \cdot n(x)) dv' \tag{2.26}$$

where M_W is some Maxwellian distribution characterizing the state of the wall and such that

$$\int_{v \cdot n(x) > 0} (v \cdot n(x)) M_W(v) dv = \int_{v \cdot n(x) < 0} |v \cdot n(x)| M_W(v) \, dv = 1,$$

which expresses the conservation of mass at the boundary.

At the macroscopic level, one can obtain two types of behaviours at the boundary : either a braking (represented by the *Dirichlet boundary condition*) or a slipping (represented by the *Navier boundary condition*), or a combination of these two phenomena (expressed by some *mixed Robin boundary condition*). This behaviour will depend of course on the nature of the boundary, but also on the viscosity of the fluid.

If the fluid is viscous, one can characterize the fluid/boundary interaction in terms of the ratio α/Ma (full braking if $\alpha/\text{Ma} \to +\infty$, perfect slipping if $\alpha/\text{Ma} \to 0$).

If the fluid is inviscid, the braking condition is not mathematically admissible. This means that the flow inside the domain will not depend (at least formally) on the nature of the boundary. The fluid/boundary interaction appears only on a thin layer (of size $1/\sqrt{\text{Re}}$), called Prandtl layer. Nevertheless this layer is generally unstable (see [61] for instance) and may give rise to turbulent effects (reflected back inside the domain).

2.3 Mathematical Theories for the Boltzmann Equation

In this section, we will introduce the main existing mathematical frameworks dealing with the Cauchy problem for the Boltzmann equation, which can be useful for the study of hydrodynamic limits. In particular, we will discuss neither the numerous results concerning the spatially homogeneous Boltzmann equation, nor the local existence results.

Let us first describe briefly the most apparent problems in trying to construct a general, good theory. In the full, general situation, known a priori estimates for the Boltzmann equation are only those which are associated to the basic physical laws, namely the formal conservation of mass and energy, and the bounds on entropy and entropy dissipation. Note that, when the physical space is unbounded, the dispersive properties of the free transport operator allow to further expect some control on the moments with respect to x-variables. Yet the Boltzmann collision integral is a quadratic operator that is purely local in the position and time variables, meaning that it acts as a convolution in the v variable, but as a pointwise multiplication in the t and x variables : thus, with the only a priori estimates which seem to hold in full generality, the collision integral is even not a well-defined distribution with respect to x-variables. This major obstruction is one of the reasons why the Cauchy problem for the Boltzmann equation is so tricky, another reason being the intricate nature of the Boltzmann operator.

2.3.1 Perturbative Framework : Global Existence of Smooth Solutions

Historically the first global existence result for the spatially inhomogeneous Boltzmann equation is due to Ukai [103], who considered initial data that are fluctuations around a global equilibrium, for instance around the reduced centered Gaussian M :

$$f_{in} = M(1 + g_{in}).$$

He proved the global existence of a solution to the Cauchy problem for (2.6) under the assumption that this initial perturbation g_{in} is smooth and small enough in a norm that involves derivatives and weights so as to ensure decay for large v.

The convenient functional space to be considered is indeed

$$H_{l,k} = \{g \equiv g(x,v) \,|\, \|g\|_{l,k} = \sup_v (1 + |v|^k) \|M^{1/2} g(\cdot, v)\|_{H^l_x} < +\infty\}.$$

Theorem 2.3.1 *Assume that the collision kernel satisfies Grad's cutoff assumption (2.8) for some $\beta \in [0,1]$. Let $g_{in} \in H_{l,k}$ for $l > 3/2$ and $k > 5/2$ such that*

$$\|g_{in}\|_{l,k} \le a_0 \tag{2.27}$$

for some a_0 sufficiently small.

Then, there exists a unique global solution $f = M(1 + g)$ with $g \in L^\infty(\mathbf{R}^+, H_{l,k}) \cap C(\mathbf{R}^+, H_{l,k})$ to the Boltzmann equation (2.6) with initial data

$$g_{|t=0} = g_{in}.$$

Remark 2.3.2 *The classical theory of the Boltzmann equation close to equilibrium, started with the works of Ukai, has been developed in the framework of hard potentials. Many such existence results, based on linearization and spectral estimates, have been proved, considering initial data which are small and very smooth perturbations of a global (Maxwellian) equilibrium.*

Using some "nonlinear energy method" instead of the spectral study of the linearized problem, and the decomposition of the solution into a "hydrodynamic" part and a "purely kinetic" part, Guo [62] was then able to extend the theory of Boltzmann's equation close to equilibrium, to cover basically all the physically meaningful range of decays of the cross-section.

Sketch of proof of Theorem 2.3.1. Such a global existence result is based on Duhamel's formula and on Picard's fixed point theorem. It requires a very precise study of the *linearized collision operator* $\mathcal{L}_M$ defined by

$$\mathcal{L}_M g = -\frac{2}{M} Q(M, Mg),$$

and more precisely of the semi-group U generated by

$$\frac{1}{\mathrm{St}} v \cdot \nabla_x + \frac{1}{\mathrm{StKn}} \mathcal{L}_M.$$

• The first step consists actually in reducing the Boltzmann equation to the integral equation

$$g = N[g], \tag{2.28}$$

where the functional N is defined by

$$\begin{aligned} N[g](t) &= U(t)g_{in} + \psi[g,g](t), \\ \psi[g,g](t) &= \frac{1}{\mathrm{Kn}} \int_0^t U(t-s) \frac{1}{M} Q(Mg, Mg)(s) ds. \end{aligned} \tag{2.29}$$

The global well-posedness of the Cauchy problem for (2.6) will then be established by proving that N is a contraction in a ball of $L^\infty(\mathbf{R}^+, H_{l,k}) \cap C(\mathbf{R}^+, H_{l,k})$.

• The second step is to prove the continuity of the linear semi-group U. Using its spectral representation and spectral estimates due to Ellis and Pinsky [45], one obtains the continuity of U in $H^l_x(L^2(Mdv))$.

In order to obtain refined estimates, and especially to gain integrability with respect to the v-variable, one has to use more about the structure of the linearized collision operator, namely the following decomposition due to Hilbert [65] (see also section 2 in Chapter 3)

$$\mathcal{L}_M = \nu - \mathcal{K}$$

where the frequency part satisfies the lower bound

$$\nu(|v|) \geq \nu_- > 0\,,$$

and the integral part $\mathcal{K}$ improves integrability in the v variable (as proved by Caflisch [23]) :

$$\mathcal{K} : H^l_x(L^2_v) \to H_{l,0}\,, \text{ and } \mathcal{K} : H_{l,k} \to H_{l,k+1}\,.$$

From the explicit formula for the semi-group $\bar{U}$ generated by

$$\frac{1}{\text{St}} v \cdot \nabla_x + \frac{1}{\text{StKn}} \nu$$

and Duhamel's formula

$$U(t) = \bar{U}(t) + \frac{1}{\text{StKn}} \int_0^t \bar{U}(t-s)\mathcal{K}U(s)ds\,,$$

we deduce that

$$\|U(t)g\|_Y \leq \exp\left(-\frac{\nu_-}{\text{StKn}}t\right)\|g\|_Y + C_{X \to Y} \int_0^t \exp\left(-\frac{\nu_-}{\text{StKn}}(t-s)\right)\|U(s)u\|_X ds$$

where $\mathcal{K}$ maps X into Y.

Iterating the process shows that, if $k > \frac{3}{2}$, there exists a nonnegative constant C_1 (depending on l and k) such that

$$\|U(t)g\|_{l,k} \leq C_1\|g\|_{l,k}\,.$$

• The continuity of the bilinear operator ψ is obtained in a very similar way.

Standard continuity estimates for Q shows that

$$\left\|\nu^{-1}\frac{1}{M}Q(Mg, Mh)\right\|_{l,k} + \left\|\frac{1}{M}Q(Mg, Mh)\right\|_{L^1(dx,(L^2(Mdv))} \leq C\|g\|_{l,k}\|h\|_{l,k}\,.$$

for $k > 3/2, l > 3/2$.

Then, starting from the spectral estimates on U and using Hilbert's decomposition to gain integrability in the v variable as previously, we obtain the expected continuity property, namely

$$\|\psi[g,h]\|_{l,k} \leq C_2 \|g\|_{l,k} \|h\|_{l,k},$$

where C_2 is a nonnegative constant depending on l and k, provided that $H_{l,k} \subset \nu H_x^l(L^2(Mdv))$, or equivalently $k > 5/2$.

- Equipped with these preliminary results, we get immediately the global existence of a unique solution to (2.6). Indeed, we have

$$\|N[g]\|_{l,k} \leq C_1 \|g_{in}\|_{l,k} + C_2 \|g\|_{l,k}^2$$

and

$$\|N[g] - N[h]\|_{l,k} \leq C_2 \left(\|g\|_{l,k} + \|h\|_{l,k} \right) \|g - h\|_{l,k}$$

Choosing a_0 and a_1 such that

$$2C_2 a_1 < 1 \text{ and } C_1 a_0 + C_2 a_1^2 \leq a_1,$$

we get that N is a contraction on the ball of radius a_1 as soon as

$$\|g_{in}\|_{l,k} \leq a_0,$$

We then conclude by Picard's fixed point theorem.   □

The first disadvantage inherent to that strategy is the need for a deep result of spectral theory. In particular, this approach fails to provide a real understanding of the coupling between relaxation and hydrodynamic modes in the full nonlinear Boltzmann equation.

For the purpose of deriving incompressible hydrodynamic limits, it would seem that Ukai's result is exactly what is needed. The difficulty is that it cannot be used as a black box, because of the potential lack of uniformity with respect to the Knudsen number Kn on the critical size of the initial perturbation that guarantees global existence. Let us mention however that Bardos and Ukai [7] have obtained the first mathematical derivation of the incompressible Navier-Stokes equations in that framework.

Nevertheless one cannot expect to extend such a result to classes of initial data with less regularity.

2.3.2 Physical Framework : Global Existence of Renormalized Solutions

For those reasons, we will use a global existence theory for the Boltzmann equation that holds for physically admissible initial data of arbitrary sizes. This theory goes back to the late 80s and is due to DiPerna and Lions [44]. For the sake of completeness, we shall sketch here the main arguments leading to that result, most of which will be detailed in the next chapter since they are also fundamental tools to study hydrodynamic limits.

Our presentation of the subject incorporates later developments of the theory of renormalized solutions :

- we will indeed consider solutions of the Boltzmann equation that converge at infinity to some uniform Maxwellian, for instance the reduced centered Gaussian M (following Lions in [73]);

- we will further present a simplification of the original proof based on compactness properties of the gain term in the collision operator (established by Lions in [72]);

- we will moreover give a weak version of the global conservation of energy and of the local conservation of momentum, involving some defect measure which characterizes the possible loss of energy at large velocities in the approximation scheme (introduced by Lions and Masmoudi in [75]);

- we will also take into account the boundary effects (modelized by Maxwell's boundary condition) (using some refined results of functional analysis due to Mischler [84][85]).

The DiPerna-Lions theory does not yield solutions that are known to solve the Boltzmann equation in the usual weak sense. Rather, it gives the existence of a global weak solution to a class of formally equivalent initial-value problems.

Definition 2.3.3 *A renormalized solution of the Boltzmann equation(2.6) (2.23) relatively to the global equilibrium M is a function*

$$f \in C(\mathbf{R}^+, L^1_{loc}(\Omega \times \mathbf{R}^3))$$

which satisfies in the sense of distributions

$$M\left(\mathrm{St}\partial_t + v \cdot \nabla_x\right)\Gamma\left(\frac{f}{M}\right) = \frac{1}{\mathrm{Kn}}\Gamma'\left(\frac{f}{M}\right)Q(f,f) \quad on \ \mathbf{R}^+ \times \Omega \times \mathbf{R}^3,$$
$$f_{|t=0} = f_{in} \geq 0 \quad on \ \Omega \times \mathbf{R}^3,$$

$$(2.30)$$

for any $\Gamma \in C^1(\mathbf{R}^+)$ such that $|\Gamma'(z)| \leq C/\sqrt{1+z}$.

We further require that for every $\varphi \in C^1_c(\bar{\Omega} \times \mathbf{R}^3)$ and every $[t_1, t_2] \subset \mathbf{R}^+$, we have

$$\mathrm{St}\int_\Omega \int M\varphi\Gamma\left(\frac{f}{M}\right)(t_2, x, v)dvdx - \mathrm{St}\int_\Omega \int M\varphi\Gamma\left(\frac{f}{M}\right)(t_1, x, v)dvdx$$
$$- \int_{t_1}^{t_2}\int_\Omega\int M(v \cdot \nabla_x\varphi)\Gamma\left(\frac{f}{M}\right)(t, x, v)dvdxdt$$
$$+ \int_{t_1}^{t_2}\int_{\partial\Omega}\int M\varphi\Gamma\left(\frac{f}{M}\right)(t, x, v)(v.n(x))dvd\sigma_x dt$$
$$= \frac{1}{\mathrm{Kn}}\int_{t_1}^{t_2}\int_\Omega\int \varphi\Gamma'\left(\frac{f}{M}\right)Q(f,f)(t, x, v)dvdxdt$$

$$(2.31)$$

with the renormalized boundary condition

$$\Gamma\left(\frac{f_{|\Sigma_-}}{M}\right) = \Gamma\left(\frac{(1-\alpha)L(f_{|\Sigma_+}) + \alpha\sqrt{2\pi}M\int f_{|\Sigma_+}(v\cdot n(x))_+dv}{M}\right) \quad on\ \Sigma_-.$$

$$(2.32)$$

With the above definition of renormalized solution relatively to M, the following existence result holds :

Theorem 2.3.4 *Assume that the collision kernel satisfies Grad's cutoff assumption (2.8) for some $\beta \in [0,1]$. Given any initial data f_{in} satisfying*

$$H(f_{in}|M) \overset{def}{=} \int_\Omega \int \left(f_{in}\log\frac{f_{in}}{M} - f_{in} + M\right)(x,v)\ dv\ dx < +\infty, \quad (2.33)$$

there exists a renormalized solution $f \in C(\mathbf{R}^+, L^1_{loc}(\Omega \times \mathbf{R}^3))$ relatively to M to the Boltzmann equation (2.6)(2.23) with initial data f_{in}.
Moreover, f satisfies
- the continuity equation

$$St\partial_t \int fdv + \nabla_x \cdot \int fvdv = 0; \quad (2.34)$$

- the momentum equation with defect measure

$$St\partial_t \int fvdv + \nabla_x \cdot \int fv \otimes vdv + \nabla_x \cdot m = 0 \quad (2.35)$$

where m is a Radon measure on $\mathbf{R}^+ \times \Omega$ with values in the nonnegative symmetric matrices;
- the entropy inequality

$$H(f|M)(t) + \int tr(m)(t) + \frac{1}{StKn}\int_0^t \int_\Omega D(f)(s,x)dsdx$$

$$+\frac{\alpha}{St}\int_0^t \int_{\partial\Omega} E(f|M)(s,x)dsd\sigma_x \leq H(f_{in}|M) \quad (2.36)$$

where $tr(m)$ is the trace of the nonnegative symmetric matrix m, the entropy dissipation $D(f)$ is defined by (2.15) and the boundary term $E(f|M)$, referred to as the Darrozès-Guiraud information is defined by

$$E(f|M)(s,x) = \int_{v\cdot n(x)>0} \left(f\log\frac{f}{M} - f + M\right)(v\cdot n(x))dv$$

$$-\left(\int f(x,v)(v\cdot n(x))_+dv\right)\log\left(\int f(x,v)\sqrt{2\pi}(v\cdot n(x))_+dv\right) \quad (2.37)$$

$$+\left(\int f(x,v)(v\cdot n(x))_+dv\right) - \frac{1}{\sqrt{2\pi}}$$

Sketch of proof of Theorem 2.3.4. We recall here the main arguments leading to that existence result, following the presentation of Golse and the author in [58] for the convergence of the approximation scheme inside the domain Ω, and the proof of Mischler in [85] for the convergence at the boundary.

Because our goal is to point out similarities between these arguments and those used in the framework of hydrodynamic limits, we focus on the weak stability of sequences (f_n) of renormalized solutions to (2.6), and do not present the underlying approximation scheme. Note that, in any case, the parameters Kn and St are fixed.

Step 1 : weak compactness results.

We have first to obtain some *weak compactness* on (f_n) using the (physical) a priori bounds.

From the **uniform bound on the relative entropy**

$$\sup_{t \in \mathbf{R}^+} H(f_n | M)(t) \leq C,$$

we deduce by Young's inequality (see (3.4) in Chapter 3) and pointwise estimates that

$$\left(\frac{f_n}{M} \right) \text{ is bounded in } L^\infty(dt, L^1_{loc}(dx, L^1(M(1 + |v|^2)dv))),$$

$$\left(\frac{f_n}{M} \right) \text{ is weakly compact in } L^1_{loc}(dtdxdv) \tag{2.38}$$

(see Lemma 3.1.2 in Chapter 3 for a detailed proof of that statement), and

$$\frac{f_n}{M} - \frac{1}{\delta} \log \left(1 + \delta \frac{f_n}{M} \right) \to 0 \text{ in } L^\infty(\mathbf{R}^+, L^1_{loc}(dx, L^1(Mdv))) \text{ uniformly in } n \tag{2.39}$$

as $\delta \to 0$. In particular, for fixed $\delta > 0$,

$$\left(\frac{Q^-(f_n, f_n)}{1 + \delta f_n / M} \right) \text{ is weakly compact in } L^1_{loc}(dtdxdv).$$

Then, from the **uniform bound on the entropy dissipation**

$$\int_0^{+\infty} \int_\Omega D(f_n)(t, x) dx dt \leq C,$$

we deduce, using a convenient splitting of the integral according to the tail of $(f_n f_{n*})/(f'_n f'_{n*})$, that for fixed $\delta > 0$,

$$\left(\frac{Q(f_n, f_n)}{1 + \delta f_n / M} \right) \text{ is weakly compact in } L^1_{loc}(dtdxdv). \tag{2.40}$$

In particular, the sequence $\frac{1}{\delta} \log(1 + \delta \frac{f_n}{M})$ (which is uniformly bounded in $L^\infty(\mathbf{R}^+, L^2_{loc}(dx, L^2(M(1 + |v|)dv)))$ by the relative entropy bound) satisfies

$$M(\mathrm{St}\partial_t + v \cdot \nabla_x) \frac{1}{\delta} \log\left(1 + \delta \frac{f_n}{M}\right) = \frac{1}{\mathrm{Kn}} \frac{Q(f_n, f_n)}{1 + \delta f_n/M} = O(1)_{L^1_{loc}(dtdxdv)}.$$

$$(2.41)$$

By interpolation (see [74] for instance), we eventually arrive at

$$\left(\frac{1}{\delta} \log(1 + \delta \frac{f_n}{M})\right) \text{ is relatively compact in } C([0, T], w - L^2_{loc}(dxMdv)),$$

which, coupled with (2.38) and (2.39), leads to

$$f_n \rightharpoonup f \text{ weakly in } L^1_{loc}(dx, L^1(dv)) \text{ locally uniformly in } t \text{ as } n \to \infty \quad (2.42)$$

modulo extraction of a subsequence.

Step 2 : strong compactness results.

In order to take limits in the renormalized Boltzmann equation, we have further to obtain some *strong compactness*, which is the matter of the second step. The crucial idea here is to use the **velocity averaging lemma** due to Golse, Lions, Perthame and Sentis [53] (and detailed in the third section of Chapter 3), stating that the moments in v of the solution to some transport equation are more regular than the function itself.

From the uniform bound on $\frac{1}{\delta} \log(1 + \delta \frac{f_n}{M})$ and the estimate (2.41) on the transport, we deduce in particular that, for all $\varphi \in C^1(\mathbf{R}^3)$ with subquadratic growth at infinity,

$$\frac{1}{\delta} \int M \log\left(1 + \delta \frac{f_n}{M}\right) \varphi(v) dv \text{ is strongly relatively compact in } L^1_{loc}(dtdx),$$

and thus by (2.39) that

$$\int f_n \varphi(v) dv \text{ is strongly relatively compact in } L^p_{loc}(dt, L^1_{loc}(dx)). \quad (2.43)$$

This convergence statement allows to take limits in the Boltzmann collision integral, once it is renormalized by some convenient macroscopic quantity. This average renormalization is here only to guarantee that all the quantities considered are at least locally integrable. Using a variant of Egorov's Theorem (namely the **Product Limit theorem** established in [44] and recalled in Appendix A), we are actually able to establish that, modulo extraction of a subsequence, for all $\phi \in C_c(\mathbf{R}^+ \times \Omega \times \mathbf{R}^3)$

$$\int \frac{Q^{\pm}(f_n, f_n)}{1 + \int f_n dv} \phi dv \longrightarrow \int \frac{Q^{\pm}(f, f)}{1 + \int f dv} \phi dv \text{ in } L^1_{loc}(dtdx). \quad (2.44)$$

Step 3 : limiting macroscopic equations.

From the previous steps, one can easily obtain the *entropy inequality* and the *variants of the conservation laws* satisfied by f.

By (2.38), we have

$$\int f_n dv \rightharpoonup \int f dv$$
$$\int f_n v dv \rightharpoonup \int f v dv$$
weakly in $L^1_{loc}(dtdx)$

which allows to take limits in the local conservation of mass. Furthermore, by the Banach-Alaoglu theorem, up to extraction of a subsequence, for each i, j

$$\int f_n v_i v_j dv \rightharpoonup \mu_{ij} \quad \text{weakly-* in } L^\infty(\mathbf{R}^+, \mathcal{M}(\Omega)). \tag{2.45}$$

By monotone convergence, one can then prove that

$$\mu_{ij} = \int f v_i v_j dv + m_{ij},$$

where m_{ij} is a nonnegative symmetric element of $L^\infty(\mathbf{R}^+, \mathcal{M}(\Omega, M_3(\mathbf{R})))$. Taking limits in the local conservation of momentum leads then to (2.35).

By weak limits

$$f_n \rightharpoonup f \text{ in } L^1_{loc}(dtdx, L^1((1+|v|)dv))$$

$$\int f_n |v|^2 dv \rightharpoonup \int f|v|^2 dv + \text{tr}(m) \text{ in } L^\infty(\mathbf{R}^+, \mathcal{M}(\Omega))$$

and

$$\frac{f_n f_{n*}}{1 + \delta \int f_n dv} \rightharpoonup \frac{f f_*}{1 + \delta \int f dv} \text{ in } L^1_{loc}(dtdx, L^1(Bdvdv_* d\sigma))$$

$$\frac{f'_n f'_{n*}}{1 + \delta \int f_n dv} \rightharpoonup \frac{f' f'_*}{1 + \delta \int f dv} \text{ in } L^1_{loc}(dtdx, L^1(Bdvdv_* d\sigma))$$

(obtained similarly as (2.44)), using the **convexity** of the functionals defining the relative entropy and the entropy dissipation, we get

$$H(f|M)(t) + \int \text{tr}(m)(t) \le \liminf_{n \to \infty} H(f_n|M)(t),$$
$$\int_0^t \int D(f)(s, x)dxds \le \liminf_{n \to \infty} \int_0^t \int D(f_n)(s, x)dxds \tag{2.46}$$

thus passing to the limit in the entropy inequality leads to (2.36) in the absence of boundary.

Step 4 : limiting renormalized kinetic equation.

The most technical step of the proof is then to take limits in *the renormalized equation* (2.30). With the information at our disposal, and although the previous step provides useful information on the nonlinear term, this convergence is not trivial, in particular because the only source of compactness in the problem, i.e. velocity averaging, does not give any information on the distribution f_n itself.

• Using pointwise estimates on $\Gamma_\delta(z) = \frac{z}{1+\delta z}$ and on its derivative, we deduce from the weak compactness statements established in Step 1 that, for all $\delta > 0$,

$$
M\Gamma_\delta\left(\frac{f_n}{M}\right) \rightharpoonup f_\delta \text{ weakly-* in } L^\infty(\mathbf{R}^+ \times \Omega \times \mathbf{R}^3),
$$

$$
\Gamma'_\delta\left(\frac{f_n}{M}\right) Q^\pm(f_n, f_n) \rightharpoonup Q_\delta^\pm \text{ weakly in } L^1_{loc}(\mathbf{R}^+ \times \Omega \times \mathbf{R}^3). \tag{2.47}
$$

Furthermore, from the relative entropy bound and the uniform convergence (2.42) we deduce that

$$
\frac{f_n}{M}(t) - \Gamma_\delta'\left(\frac{f_n}{M}\right)(t) \to 0 \text{ in } L^1_{loc}(dx, L^1(dv)) \text{ uniformly in } t, n \text{ as } \delta \to 0,
$$

and thus that

$$
f_\delta \to f \text{ as } \delta \to 0 \text{ in } L^1_{loc}(dx, L^1(dv)) \text{ uniformly in } t, \\
\text{and a.e. on } \mathbf{R}^+ \times \Omega \times \mathbf{R}^3. \tag{2.48}
$$

The idea is then to take limits in

$$
M(\mathrm{St}\partial_t + v \cdot \nabla_x) \log\left(1 + \frac{f_\delta}{M}\right) = \frac{1}{\mathrm{Kn}} \frac{Q_\delta^+ - Q_\delta^-}{1 + f_\delta/M},
$$
$$
f_{\delta|t=0} = \Gamma_\delta(f_{in}).
$$

• By the strong compactness statements (2.43) established in Step 2, and the Product Limit theorem, we have, for all $\delta > 0$,

$$
\frac{Q^-(f_n, f_n)}{(1 + \delta f_n/M)^2} = \frac{f_n}{(1 + \delta f_n/M)^2} \iint f_{n*} B dv_* d\sigma
$$
$$
\rightharpoonup \tilde{f}_\delta \iint f_* B dv_* d\sigma \text{ in } L^1_{loc}(\mathbf{R}^+ \times \Omega \times \mathbf{R}^3)
$$

with $\tilde{f}_\delta \le f_\delta$ and

$$
\tilde{f}_\delta \to f \text{ as } \delta \to 0 \text{ in } L^1_{loc}(dx, L^1(dv)) \text{ uniformly in } t, \\
\text{and a.e. on } \mathbf{R}^+ \times \Omega \times \mathbf{R}^3,
$$

using the same arguments as for (2.48). We then obtain the *convergence of the loss term* (up to extraction of a subsequence)

$$
\frac{Q_\delta^-}{1 + f_\delta/M} \to \frac{f}{1 + f/M} \iint f_* B dv_* d\sigma \text{ as } \delta \to 0 \text{ a.e. on } \mathbf{R}^+ \times \Omega \times \mathbf{R}^3 \tag{2.49}
$$

and thus in $L^1_{loc}(\mathbf{R}^+ \times \Omega \times \mathbf{R}^3)$ by Lebesgue's theorem.

• The *convergence of the gain term* is more complicated to establish. Starting from

$$\Gamma'_\delta\left(\frac{f_n}{M}\right)\frac{Q^+(f_n, f_n)}{1 + \int f_n dv} \leq \left(\frac{f_n}{M}\right)\frac{Q^+(f_n, f_n)}{1 + \int f_n dv}$$

then integrating against some $\phi = \phi(v) \geq 0$ and taking limits as $n \to \infty$, we get

$$Q^+_\delta \leq Q^+(f, f) \quad \text{a.e. on } \mathbf{R}^+ \times \Omega \times \mathbf{R}^3$$

using the convergence (2.44) obtained in Step 2, and the Product Limit theorem.

Then, introducing some suitable decomposition according to the tail of $(f'_n f'_{n*})/(f_n f_{n*})$, and using the convergence (2.44) and the Product Limit theorem, we establish that, for all $\lambda > 0$,

$$\frac{Q^+(f_n, f_n)}{1 + \lambda \int f_{n*} dv_*} \rightharpoonup \frac{Q^+(f, f)}{1 + \lambda \int f_* dv_*} \quad \text{weakly in } L^1_{loc}(\mathbf{R}^+ \times \Omega \times \mathbf{R}^3).$$

Starting from a refined decomposition, and using the convergence of the entropy dissipation in the vague sense of measures, we then obtain that, for all $\lambda > 0$,

$$\frac{Q^+(f, f)}{1 + \lambda \int f_* dv_*} \leq \liminf_{\delta \to 0} Q^+_\delta.$$

Finally, we get

$$\frac{Q^+_\delta}{1 + f_\delta/M} \to \frac{Q^+(f, f)}{1 + f/M} \quad \text{as } \delta \to 0 \text{ a.e. on } \mathbf{R}^+ \times \Omega \times \mathbf{R}^3 \qquad (2.50)$$

and thus in $L^1_{loc}(\mathbf{R}^+ \times \Omega \times \mathbf{R}^3)$ by Lebesgue's theorem.

• Combining all results leads to

$$M(\text{St}\partial_t + v \cdot \nabla_x) \log\left(1 + \frac{f}{M}\right) = \frac{1}{\text{Kn}} \frac{Q^+(f, f) - Q^-(f, f)}{1 + f/M},$$

with initial condition $f|_{t=0} = f_{in}$ (since the convergence is uniform in t). It remains then to check that the same identity holds for any admissible renormalization Γ

$$M(\text{St}\partial_t + v \cdot \nabla_x)\Gamma\left(\frac{f}{M}\right) = \frac{1}{\text{Kn}}\Gamma'\left(\frac{f}{M}\right)(Q^+(f, f) - Q^-(f, f)), \qquad (2.51)$$

which is done by composition if $|\Gamma'(z)| \leq C(1 + z)^{-1}$, and else by approximation, using the fact that $Q(f, f)/\sqrt{1 + f/M}$ is controlled by the entropy dissipation and the relative entropy (see the proof of Proposition 4.3.1 in Chapter 4 for an analogous result).

Step 5 : limiting boundary conditions.

In the case of a spatial domain with boundary, it remains then to take limits in *Maxwell's boundary condition*. This requires powerful tools of functional analysis, which are consequences of **Chacon's Biting Lemma** and are stated in Appendix C.

Let us first note that the boundary term obtained formally in the entropy inequality

$$\int_0^t \int_{\partial\Omega} \int \left(f_{n|\Sigma^+} \log \frac{f_{n|\Sigma^+}}{M} - f_{n|\Sigma^+} + M \right) (s,x,v)(v \cdot n(x))_+ dv d\sigma_x ds$$
$$- \int_0^t \int_{\partial\Omega} \int \left(f_{n|\Sigma^-} \log \frac{f_{n|\Sigma^-}}{M} - f_{n|\Sigma^-} + M \right) (s,x,R_x v)(v \cdot n(x))_+ dv d\sigma_x ds$$

controls the Darrozès-Guiraud information

$$\alpha \int_0^t \int_{\partial\Omega} E(f_n)(s,x) d\sigma_x ds$$

defined by (2.37) (by a simple convexity argument), and so we start from a sequence (f_n) such that the Darrozès-Guiraud information $E(f_n)$ is uniformly bounded in $L^1(\mathbf{R}^+ \times \partial\Omega)$.

The trace is then defined by some Green's formula written on the renormalized equation. The main difficulty to take limits in the renormalized form (2.32) of Maxwell's boundary condition, is therefore the lack of an a priori bound on the trace, giving in particular some local equi-integrability in v.

- We first establish the following *renormalized convergence* (see Appendix C for a precise definition of this notion)

$$f_{n|\partial\Omega} \to f_{\partial\Omega} \text{ in renormalized sense on } \mathbf{R}^+ \times \partial\Omega \times \mathbf{R}^3, \tag{2.52}$$

using the a priori estimates coming from the inside, and the weak formulation (2.31) of the renormalized Boltzmann equation.

Starting from (2.31) with

$$\varphi(x,v) = \frac{v \cdot n(x)}{1 + |v|^2} \chi(x)$$

where $\chi \in C_c^\infty(\mathbf{R}^3, \mathbf{R}^+)$ and n denotes some vector field of $W^{1,\infty}(\bar\Omega)$ which coincides with the outward unit normal vector at the boundary, we get

$$\int_{t_1}^{t_2} \int_{\partial\Omega} \int M \frac{(v \cdot n(x))^2 \chi(x)}{1 + |v|^2} \Gamma\left(\frac{f_{n|\partial\Omega}}{M}\right)(t,x,v) dv d\sigma_x dt$$
$$= \text{St} \int_\Omega \int M \frac{(v \cdot n(x))\chi(x)}{1 + |v|^2} \Gamma\left(\frac{f_n}{M}\right)(t_1,x,v) dv dx$$
$$- \text{St} \int_\Omega \int M \frac{(v \cdot n(x))\chi(x)}{1 + |v|^2} \Gamma\left(\frac{f_n}{M}\right)(t_2,x,v) dv dx \tag{2.53}$$
$$+ \int_{t_1}^{t_2} \int_\Omega \int M(v \cdot \nabla_x) \left(\frac{(v \cdot n(x))\chi(x)}{1 + |v|^2}\right) \Gamma\left(\frac{f_n}{M}\right)(t,x,v) dv dx dt$$
$$+ \frac{1}{\text{Kn}} \int_{t_1}^{t_2} \int_\Omega \int \frac{(v \cdot n(x))\chi(x)}{1 + |v|^2} \Gamma'\left(\frac{f_n}{M}\right) Q(f_n,f_n)(t,x,v) dv dx dt.$$

Thus from the uniform bounds obtained in Step 1 and Cauchy-Schwarz inequality, we deduce that

$$M\Gamma\left(\frac{f_{n|\partial\Omega}}{M}\right) \text{ is weakly compact in } L^1_{loc}(dtd\sigma_x, L^1(|v\cdot n(x)|dv)).$$

Using the convergence results stated in Step 4, we can take limits as $n\to\infty$ in the right-hand side of (2.53), and then identify the limit writing Green's formula for the limiting kinetic equation (2.51). We thus obtain

$$M\Gamma\left(\frac{f_{n|\partial\Omega}}{M}\right) \rightharpoonup M\Gamma\left(\frac{f_{|\partial\Omega}}{M}\right) \text{ weakly in } L^1_{loc}(dtd\sigma_x, L^1(|v\cdot n(x)|dv)).$$

which implies the renormalized convergence (2.52). Note that, up to extraction of a subsequence, we also get the pointwise convergence

$$f_{n|\partial\Omega} \to f_{|\partial\Omega} \text{ a.e. on } \mathbf{R}^+\times\partial\Omega\times\mathbf{R}^3.$$

• Then, using the uniform bound on the *Darrozès-Guiraud information*, we prove that

$$\int f_{n|\partial\Omega}(v\cdot n(x))_+dv \to \tilde{f}_{|\partial\Omega} \text{ in renormalized sense on } \mathbf{R}^+\times\partial\Omega, \quad (2.54)$$

for some measurable, almost everywhere finite function $\tilde{f}_{|\partial\Omega}$.

Indeed, remarking that

$$(z\log z - z + 1) - (y\log y - y + 1) - (z-y)\log y = \int_0^1 \frac{|z-y|^2}{\tau x + (1-\tau)y}d\tau$$
$$\geq \left(\sqrt{z} - \sqrt{y}\right)^2$$

and that

$$\int\left(f_n - \sqrt{2\pi}M\int f_n(v\cdot n(x))_+dv\right)\log\left(\int f_n(v\cdot n(x))_+dv\right)(v\cdot n(x))_+dv = 0$$

we get

$$\int\left(\sqrt{\frac{f_n}{M}} - \sqrt{\int f_n\sqrt{2\pi}(v\cdot n(x))_+dv}\right)^2 M(v\cdot n(x))_+dv \leq 2E(f_n|M)$$

which, coupled with the uniform bound on the Darrozès-Guiraud information, shows that

$$\sqrt{\frac{f_n}{M}} - \sqrt{\int f_n\sqrt{2\pi}(v\cdot n(x))_+dv} \text{ is weakly compact in } L^2(dt(v\cdot n(x))_+d\sigma_x Mdv),$$

$$(2.55)$$

and thus converges a.e. on Σ_+ up to extraction of a subsequence.

Therefore, from the decomposition

$$\sqrt{\int f_n \sqrt{2\pi}(v \cdot n(x))_+ dv} = \sqrt{\int f_n \sqrt{2\pi}(v \cdot n(x))_+ dv - \sqrt{\frac{f_n}{M}}} + \sqrt{\frac{f_n}{M}}$$

the renormalized convergence (2.52) and the weak compactness (2.55), we deduce that (2.54) holds up to extraction of a subsequence.

- It remains then to characterize the limit $\tilde{f}_{|\partial\Omega}$ in terms of $f_{|\partial\Omega}$, which requires a variant of Chacon's Biting Lemma giving some partial equiintegrability on $f_{n|\partial\Omega}$ with respect to the v variables.

From (2.54) and the uniform bound on the Darrozès-Guiraud information, we deduce by Proposition C.4 in Appendix that for every $\varepsilon > 0$ and every compact $K \subset \mathbf{R}^+ \times \partial\Omega$, one can find some $A \subset K$ with

$$\int_{K \backslash A} dt d\sigma_x < \varepsilon \text{ and } f_{n|\partial\Omega} \rightharpoonup f_{|\partial\Omega} \text{ weakly in } L^1(A \times \mathbf{R}^3, dt(v \cdot n(x))_+ d\sigma_x dv).$$

In particular,

$$\tilde{f}_{|\partial\Omega} = \int f_{|\partial\Omega}(v \cdot n(x))_+ dv \text{ on every such } A,$$

and thus a.e. on $\mathbf{R}^+ \times \partial\Omega$.

We are then able to take limits in the renormalized form of Maxwell's boundary condition (2.32), which leads to

$$\Gamma\left(\frac{f_{|\Sigma_-}}{M}\right) = \Gamma\left(\frac{(1-\alpha)L(f_{|\Sigma_+}) + \alpha K(f_{|\Sigma_+})}{M}\right) \quad \text{on } \Sigma_-.$$

Furthermore, using the convexity of the Darrozès-Guiraud information (also established in Proposition C.4), we get

$$\int_0^t \int_{\partial\Omega} E(f|M)(s,x) d\sigma_x ds \leq \liminf_{n \to \infty} \int_0^t \int_{\partial\Omega} E(f_n|M)(s,x) d\sigma_x ds,$$

which concludes the proof of the entropy inequality (2.36) studied in Step 3, in the case of a spatial domain with boundary. $\qquad\Box$

2.3.3 Further Results in One Space Dimension

In the one spatial dimensional case, the previous result has actually been improved by Cercignani [29], who established the global existence of weak solutions to (2.6) satisfying in particular the global conservation of energy.

The key idea of that theory is to introduce the weak form of the collision term, and the corresponding suitable notion of weak solution. For the sake of simplicity, we will restrict our attention to the case of a spatial domain without boundary, for instance the periodic box $\mathbf{T}$.

Definition 2.3.5 *A weak solution to the one-dimensional Boltzmann equation (2.6) is a function*

$$f \in C(\mathbf{R}^+, L^1(\mathbf{T} \times \mathbf{R}^3))$$

such that, for every test function $\varphi \in C_c^1(\mathbf{R}^+ \times \mathbf{T} \times \mathbf{R}^3)$ which is twice differentiable as a function of v with second derivatives uniformly bounded with respect to x and t, we have

$$\iiint f(\mathrm{St}\partial_t\varphi + v_1\partial_x\varphi)(t,x,v)dxdvdt + \iint f_{in}\varphi(0,x,v)dxdv$$
$$= \frac{1}{2\mathrm{Kn}} \iint \left(\iiint ff_*(\varphi + \varphi_* - \varphi' - \varphi'_*)B(v - v_*, \sigma)dvdv_*d\sigma \right)(t,x)dxdt \tag{2.56}$$

With the above definition of weak solution, the following existence result holds :

Theorem 2.3.6 *Assume that the collision kernel B is bounded and satisfies Grad's cutoff assumption (2.8) as well as*

$$\int_{S^2} (1 + \cos\theta)B(v - v_*, \sigma)d\sigma \geq r \int_{S^2} B(v - v_*, \sigma)d\sigma \tag{2.57}$$

for some $r > 0$. Given any initial data $f_{in} \in L^1_{loc}(\mathbf{T} \times \mathbf{R}^3)$ satisfying

$$H(f_{in}|M) \stackrel{def}{=} \iint \left(f_{in} \log \frac{f_{in}}{M} - f_{in} + M \right)(x,v) \, dv \, dx < +\infty, \tag{2.58}$$

there exists a weak solution $f \in C(\mathbf{R}^+, L^1(\mathbf{T} \times \mathbf{R}^3))$ to (2.6) with initial data f_{in}.

Furthermore this solution satisfies the continuity equation (2.34), the momentum equation (2.35) without defect measure and the entropy inequality (2.36) without defect measure, as well as the energy conservation

$$\iint f(t,x,v)|v|^2 dvdx = \iint f_{in}(x,v)|v|^2 dvdx.$$

Sketch of Proof of Theorem 2.3.6. The idea is to use the knowledge that there is a renormalized solution in the sense of DiPerna-Lions, and to establish estimates which entail that this solution is indeed a weak solution in the sense defined above. As usual these estimates will be obtained by formal computations, which can be justified for approximate solutions to the Boltzmann equation (2.6), and then established for any renormalized solution by passing to the limit.

The crucial tool to establish these estimates, which is specific to the one-dimensional case, is the functional

$$I(f)(t) \stackrel{\text{def}}{=} \iint_{x<y} \iint (v_1 - v_{1*}) f(t, x, v) f(t, y, v_*) dv_* dv dx dy \qquad (2.59)$$

which extends the *potential for interaction* introduced by Bony in the one-dimensional discrete velocity context. No functional with similar pleasant properties is known, at this time, in more than one dimension. Note indeed that, because of the bounds on the total mass $\iint f(t, x, v) dv dx$ and on the total momentum $\iint f(t, x, v) v_1 dv dx$ in x-direction, we have the following control over the functional $I(f)(t)$

$$\forall t \in \mathbf{R}, \quad |I(f)(t)| \leq C_{in},$$

where C_{in} is a constant depending only on the initial data.

- The first step of the proof consists then in using that bound to establish the following basic estimates

$$\int_0^t \int \iint (v_1 - u_1(s, x))^2 f(s, x, v) f(s, x, v_*) dv_* dv dx ds \leq C_{in},$$

$$\int_0^t \int \iint |v - v_*|^2 f(s, x, v) f(s, x, v_*) B(v - v_*, \sigma) d\sigma dv_* dv dx ds \leq C_{in}$$
$$(2.60)$$

where C_{in} is as previously some constant depending only on the initial data, and u_1 is the bulk velocity defined by

$$u_1(s, x) = \frac{\int v_1 f(s, x, v) dv}{\int f(s, x, v) dv}.$$

A short calculation with proper use of the collision invariants of the Boltzmann collision operator shows that

$$I(f)(t) - I(f)(0) = -\int_0^t \int \iint (v_1 - v_{1*})^2 f(s, x, v) f(s, x, v_*) dv_* dv dx dt.$$

which immediately gives the first estimate in (2.60), remarking that

$$\int (v_1 - v_{1*})^2 f(s, x, v_*) dv_* \geq \int (v_1 - u_1)^2 f(s, x, v_*) dv_*.$$

From the weak form of the Boltzmann equation, we deduce using the conservation of mass and momentum that

$$2\mathrm{Kn} \iint f v_1^2 dv dx - 2\mathrm{Kn} \iint f_{in} v_1^2 dv dx =$$
$$\int_0^t \iiiint f f_* ((v_1 - u_1)^2 + (v_{1*} - u_1)^2 - (v_1' - u_1)^2 - (v_{1*}' - u_1)^2) B dv dv_* d\sigma dx dt$$

The loss term is bounded because of the bound on the collision frequency and the first estimate in (2.60), and the left hand side is bounded because of the energy bound. We therefore deduce that the gain term is also bounded. Then by explicit computations based on symmetries and assumption (2.57), we get the second estimate in (2.60).

- Equipped with these preliminary estimates, we are now able to prove that the integral defining the weak form of the collision operator is bounded in terms of constants depending on the initial data, for any test function $\varphi \equiv \varphi(t, x, v)$ which is twice differentiable as a function of v with second derivatives uniformly bounded with respect to x and t.

The result follows from Taylor's formula at second order, remarking that the expression multiplying the first derivatives is zero because of momentum conservation. We indeed have, using the second estimate in (2.60),

$$\iint \left(\iiint f f_* |\varphi + \varphi_* - \varphi' - \varphi'_*| B(v - v_*, \sigma) dv dv_* d\sigma \right) dx dt$$

$$\leq C \iiiint f f_* \left(|v - v_*|^2 + |v - v'|^2 + |v - v'_*|^2 \right) B(v - v_*, \sigma) dv dv_* d\sigma dx dt$$

$$\leq 6C \iiiint f f_* |v - v_*|^2 B(v - v_*, \sigma) dv dv_* d\sigma dx dt \leq 6CC_{in}$$

which shows that the weak form of $Q(f, f)$ is well-defined.   □

Remark 2.3.7 *The present result can actually be extended to slightly more general situations.*

- *Easy modifications, presented for instance in the paper [30] by Cercignani and Illner, allow to deal with the case of different boundary conditions, namely to consider the case of a slab with diffusive boundary conditions.*
- *Let us now discuss the assumptions on the collision kernel B. In principle, solutions for inverse power potentials might be considered without introducing Grad's cutoff (2.8) : this would require considering approximate solutions of the Boltzmann equation without cutoff, and study precisely the convergences, instead of using the knowledge that there is a renormalized solution.*

On the other hand, the present version of the result does not allow a growth for large values of the relative velocity $|v - v_*|$, i.e. excludes hard spheres and potentials harder than the inverse fifth power. This is an important simplification, which perhaps might be removed by much harder work.

This particular structure of the Boltzmann equation in one space dimension is reminiscent of the specificity of the one dimensional hyperbolic systems of conservation laws. In particular the functional referred to as the potential for interaction, and obtained by doubling the space variable, has to be compared with Glimm's functional for systems of conservation laws, which could be a track to investigate the compressible hydrodynamic limits.

3

Mathematical Tools for the Derivation of Hydrodynamic Limits

In all existing works on the subject, the general strategy to derive hydrodynamic limits is to proceed by analogy, that is to recognize the structure of the expected limiting hydrodynamic model in the corresponding scaled Boltzmann equation. This explains for instance why all hydrodynamic limits are not equally understood.

   The aim of this chapter is therefore to detail these analogies, focusing our attention on the point of view of functional analysis.

3.1 Physical a Priori Estimates : Definition of Suitable Functional Spaces

Let us first recall from the previous chapter, that, in a general setting, the only a priori bounds we dispose of for the solutions to the scaled Boltzmann equation are those coming from physics, namely from the global conservation of mass and energy, and from Boltzmann's H theorem. Considering the case of a gas which is at Maxwellian equilibrium M at infinity (or which is confined in a domain with diffuse reflection according to the distribution M at the boundary), all these physical estimates lead to the unique scaled relative entropy inequality

$$\iint Mh\left(\frac{f-M}{M}\right)(t,x,v)dxdv + \frac{1}{\text{KnSt}}\int_0^t\int D(f)(s,x)dxds$$
$$+\frac{\alpha}{\text{St}}\int_0^t\int_{\partial\Omega} E(f|M)(s,x)dxds$$
$$\leq \iint Mh\left(\frac{f_{in}-M}{M}\right)(x,v)dxdv \overset{\text{def}}{=} H_{in}.$$

$$(3.1)$$

where h is the convex function defined on $]-1,+\infty[$ by

$$h(z) = (1+z)\log(1+z) - z,$$

L. Saint-Raymond, *Hydrodynamic Limits of the Boltzmann Equation*,
Lecture Notes in Mathematics 1971, DOI: 10.1007/978-3-540-92847-8_3,
© Springer-Verlag Berlin Heidelberg 2009

and D and E are respectively the entropy dissipation defined by (2.15) and the Darrozès-Guiraud information defined by (2.37).

We thus expect the appropriate functional framework for the study of hydrodynamic limits to be defined in terms of these three functionals, namely the relative entropy, the entropy dissipation, and the Darrozès-Guiraud information.

3.1.1 The Entropy Bound

The relative entropy bound

$$\forall t \geq 0, \quad H(f|M)(t) \stackrel{\text{def}}{=} \iint Mh\left(\frac{f-M}{M}\right)(t,x,v)dxdv \leq H_{in}$$

controls the distance of the distribution function f to the background Maxwellian M, in a sense to be precised.

In the General (Non Perturbative) Framework,

that is in the situation when

$$H_{in} = O(1),$$

the entropy bound leads to the following macroscopic bounds :

Lemma 3.1.1 *Let f be a (nonnegative) distribution function such that*

$$\iint Mh\left(\frac{f-M}{M}\right)(t,x,v)dxdv \leq H_{in},$$

where M is the global Maxwellian of density R_0, bulk velocity U_0 and temperature T_0.

Then, the corresponding macroscopic fields R, U, T defined by

$$R(t,x) = \int f(t,x,v)dv, \quad RU(t,x) = \int f(t,x,v)vdv,$$
$$R(|U|^2 + 3T)(t,x) = \int f(t,x,v)|v|^2dv.$$

satisfy the bounds

$$\int R_0 h\left(\frac{R}{R_0} - 1\right)(t,x)dx \leq H_{in},$$
$$\frac{1}{2T_0}\int R|U - U_0|^2(t,x)dx \leq H_{in},$$
$$\frac{3}{2}\int R\left(\frac{T}{T_0} - \log\frac{T}{T_0} - 1\right)(t,x)dx \leq H_{in}.$$

Proof. Such a result is proved very easily using the explicit formula for

$$H(\mathcal{M}_f|M)(t) \stackrel{\text{def}}{=} \iint Mh\left(\frac{\mathcal{M}_f - M}{M}\right)(t, x, v)dxdv$$

where $\mathcal{M}_f$ is the local Maxwellian of same moments as f defined by (2.21), i.e.

$$\mathcal{M}_f(t, x, v) = \frac{R(t, x)}{(2\pi T(t, x))^{3/2}} \exp\left(-\frac{|v - U(t, x)|^2}{2T(t, x)}\right),$$

and the fundamental inequality

$$H(\mathcal{M}_f|M) \le H(f|M)$$

expressing the fact that the Maxwellian distribution minimizes the entropy for fixed density, bulk velocity and temperature.

Indeed

$$H(f|M)(t) = \int \left(f \log \frac{f}{M} - f + M\right)(t, x, v)dxdv$$

$$= \int \left(f \log \frac{f}{\mathcal{M}_f} - f + \mathcal{M}_f\right)(t, x, v)dxdv$$

$$+ \int \left(\mathcal{M}_f \log \frac{\mathcal{M}_f}{M} - \mathcal{M}_f + M\right)(t, x, v)dxdv$$

$$+ \int (f - \mathcal{M}_f) \log \frac{\mathcal{M}_f}{M}(t, x, v)dxdv$$

$$= H(f|\mathcal{M}_f)(t) + H(\mathcal{M}_f|M)(t)$$

using the fact that f and $\mathcal{M}_f$ have the same moments up to order 2, and that $\log(\mathcal{M}_f/M)$ (considered as a function of v) is nothing else than a linear combination of 1, v and $|v|^2$. $\qquad\square$

This implies in particular that the functional setting for the study of the compressible Euler limit is very intricated, which is not surprising in view of the few stability results established for this system.

For Fluctuations around the Global Maxwellian State M,

that are functions of the form

$$f = M(1 + \text{Ma}\, g), \tag{3.2}$$

or more precisely for solutions of the Boltzmann equation such that

$$H_{in} = \iint Mh\left(\text{Ma}\, g_{in}\right)(x, v)dxdv = O(\text{Ma}^2),$$

the functional spaces defined asymptotically as Ma $\to 0$ by the relative entropy are flat spaces with hilbertian structure.

Without loss of generality, we will assume in all the sequel that M is the reduced centered Gaussian

$$M(v) = \frac{1}{(2\pi)^{3/2}} \exp\left(-\frac{|v|^2}{2}\right),$$

meaning that $R_0 = T_0 = 1$, and $U_0 = 0$, which can be obtained by a rescaling process and a galilean change of coordinates.

The implications of the relative entropy bound that we shall discuss here are straightforward consequences of pointwise inequalities satisfied by the non-linearity h that defines the relative entropy. Recall that

$$H(f|M) = \iint h(\mathrm{Mag})M\,dv\,dx\,;$$

since

$$h(z) \sim \frac{1}{2}z^2 \quad \text{as } z \to 0, \tag{3.3}$$

one could think that the relative entropy bound is more or less equivalent to a L^2 estimate of the type

$$\iint |g(t,x,v)|^2 M\,dv\,dx \leq 2\frac{H_{in}}{\mathrm{Ma}^2}\,.$$

However, this is not entirely correct, since g can take values that are very large compared with $1/\mathrm{Ma}$, for which replacing $h(z)$ by $\frac{1}{2}z^2$ is not justified.

Instead of (3.3), we must use global properties of h. First, h satisfies Young's inequality

$$pz \leq h^*(p) + h(z), \quad \forall p, z \geq 0,$$

where h^* is the Legendre dual of h:

$$h^*(p) = e^p - p - 1.$$

Notice that h^* is super-quadratic (as can be seen from the Taylor series that defines h^*): in other words

$$h^*(\lambda p) \leq \lambda^2 h^*(p), \quad \forall p \geq 0, \forall \lambda \in [0,1].$$

Also, notice that

$$h(|z|) \leq h(z), \quad \forall z > -1.$$

Putting all these inequalities together, we arrive at the following improvement of Young's inequality above:

$$p|z| \leq \lambda h^*(p) + \frac{1}{\lambda}h(z), \quad \forall p \geq 0, \forall z \geq -1, \forall \lambda \in (0,1]. \tag{3.4}$$

Lemma 3.1.2 *For each sequence* $\mathrm{Ma}_n \to 0$, *let* (f_n) *be a sequence of measurable, a.e. nonnegative distribution functions such that*

$$\forall t \geq 0, \quad H(f_n|M)(t) \leq C\mathrm{Ma}_n^2.$$

Then, the sequence of fluctuations (g_n) *defined by*

$$f_n = M(1 + \mathrm{Ma}_n g_n)$$

is weakly relatively compact in $L^1_{loc}(dtdx, L^1(M(1 + |v|^2)dv))$.

Proof. Pick $\eta \in (0, 1]$; Young's inequality (3.4) implies that, for each n such that $\mathrm{Ma}_n \in (0, \eta)$, i.e. for all but a finite number of n's, one has

$$(1 + |v|^2)|g_n| \leq \frac{4\eta}{\mathrm{Ma}_n^2} h(\mathrm{Ma}_n g_n) + \frac{4}{\eta} h^* \left(\frac{1 + |v|^2}{4} \right)$$

by taking

$$z = \mathrm{Ma}_n g_n, \quad p = \frac{1}{4} \frac{\mathrm{Ma}_n}{\eta} (1 + |v|^2) \text{ and } \lambda = \frac{\mathrm{Ma}_n}{\eta}.$$

Consider first the case $\eta = 1$; hence, for each measurable set $E \subset \Omega$ of finite measure

$$\int_E \int (1 + |v|^2)|g_n(t)|Mdvdx \leq 4C + 4|E| \int \exp(\frac{1}{4}(1 + |v|^2))Mdv.$$

Hence

$$(1 + |v|^2)g_n \text{ is bounded in } L^\infty(dt; L^1_{loc}(dx : L^1(Mdv))).$$

For general $\eta \in (0, 1)$, for each n such that $\mathrm{Ma}_n < \eta$

$$\int_E \int (1 + |v|^2)|g_n(t)|Mdvdx \leq 4\eta C + \frac{4}{\eta}|E| \int \exp(\frac{1}{4}(1 + |v|^2))Mdv.$$

Choosing $\eta = |E|^{1/2}$ shows that

$$\int_E \int (1 + |v|^2)|g_n(t)|Mdvdx \leq C|E|^{1/2}$$

for each n such that $\mathrm{Ma}_n < |E|^{1/2}$, i.e. for all but a finite number of n's. This shows that the sequence

$$(1 + |v|^2)g_n \text{ is uniformly integrable on } \Omega \times \mathbf{R}^3$$

uniformly in $t \geq 0$. By Dunford-Pettis' criterion, this implies the announced weak compactness for the sequence (g_n) of fluctuations. $\square$

The formal argument given at the beginning of this paragraph suggests however that, in the vanishing Mach limit Ma → 0, the limiting fluctuation belongs to $L^2(Mdvdx)$ uniformly in t. Hence the weighted L^1-bound given by Lemma 3.1.2 is certainly not optimal.

We propose therefore to consider the following *renormalized fluctuation*

$$\hat{g} = \frac{2}{\text{Ma}} \left(\sqrt{\frac{f}{M}} - 1 \right), \tag{3.5}$$

instead of the fluctuation

$$g = \frac{1}{\text{Ma}} \left(\frac{f}{M} - 1 \right).$$

The advantage of this renormalized fluctuation over the original one is explained in the next lemma.

Lemma 3.1.3 *For each sequence* $\text{Ma}_n \to 0$, *let* (f_n) *be a sequence of measurable, a.e. nonnegative distribution functions such that*

$$\forall t \geq 0, \quad H(f_n|M)(t) \leq C\text{Ma}_n^2.$$

Then, the family of renormalized fluctuations $(\hat{g}_n)$ *defined by (3.5) is bounded in* $L^\infty(\mathbf{R}_+; L^2(Mdvdx))$.

Proof. The elementary inequality

$$\frac{1}{2}h(z) \geq (\sqrt{1+z} - 1)^2, \quad \forall z > -1 \tag{3.6}$$

implies that

$$\iint \hat{g}_n^2(t, x, v)M(v)dxdv \leq \frac{2}{\text{Ma}_n^2}H(f_n|M)(t) \leq 2C,$$

which is the announced result.   □

A natural application of this refined a priori estimate is to decompose

$$g = \hat{g} + \frac{1}{4}\text{Ma}\hat{g}^2. \tag{3.7}$$

Therefore, we see that the fluctuation g is bounded in $L^2(Mdvdx)$, up to a remainder of order Ma in $L^1(Mdvdx)$, uniformly in $t \geq 0$.

3.1.2 The Darrozès-Guiraud Information

The bound on the Darrozès-Guiraud information

$$\frac{\alpha}{\mathrm{St}} \int_0^{+\infty} \int_{\partial\Omega} E(f|M)(t,x)d\sigma_x dt \leq H_{in}$$

is expected to give some control on the trace of the distribution function f on the boundary $\partial\Omega$, or more precisely on the variation (with respect to the v variable) of the trace

$$f_{|\partial\Omega} - M \int f_{|\partial\Omega}\sqrt{2\pi}(v \cdot n(x))_+ dv$$

on Σ_+. We have indeed the following estimate

Lemma 3.1.4 *Let (f_n) be a sequence of measurable, a.e. nonnegative distribution functions such that*

$$\forall t > 0, \quad \int_0^t \int_{\partial\Omega} E(f|M)(s,x)d\sigma_x ds \leq C\mathrm{Ma}_n^2 \frac{\mathrm{St}_n}{\alpha_n}.$$

Then the family of renormalized trace variations $(\hat{\eta}_n)$ defined by

$$\hat{\eta}_n = \frac{2}{\mathrm{Ma}_n}\sqrt{\frac{\alpha_n}{\mathrm{St}_n}}\mathbf{1}_{\Sigma_+}\left(\sqrt{\frac{f_{n|\partial\Omega}}{M}} - \sqrt{\int f_{n|\partial\Omega}\sqrt{2\pi}(v\cdot n(x))_+ dv}\right) \qquad (3.8)$$

is uniformly bounded in $L^2_{loc}(dt, L^2(M(v\cdot n(x))_+ d\sigma_x dv))$.

Proof. Denoting by $\langle\cdot\rangle_{\partial\Omega}$ the average of any quantity defined on the boundary

$$\langle g\rangle_{\partial\Omega} = \int Mg_{|\partial\Omega}\sqrt{2\pi}(v\cdot n(x))_+ dv,$$

we get

$$\frac{\alpha_n}{\mathrm{Ma}_n^2\mathrm{St}_n}\int_0^t \int_{\partial\Omega}\left\langle h\left(\frac{f_{n|\partial\Omega} - M}{M}\right) - h\left(\left\langle\frac{f_n - M}{M}\right\rangle_{\partial\Omega}\right)\right\rangle_{\partial\Omega} d\sigma_x ds \leq C\sqrt{2\pi}.$$

By Taylor's formula,

$$\left\langle h\left(\frac{f_n - M}{M}\right) - h\left(\left\langle\frac{f_n - M}{M}\right\rangle_{\partial\Omega}\right)\right\rangle_{\partial\Omega}$$
$$= \frac{1}{2}\int_0^1\left\langle\left(\left(\frac{f_{n|\partial\Omega}}{M}\right) - \left\langle\frac{f_n}{M}\right\rangle_{\partial\Omega}\right)^2 h''\left(\tau\frac{f_{n|\partial\Omega}}{M} + (1-\tau)\left\langle\frac{f_n}{M}\right\rangle_{\partial\Omega} - 1\right)\right\rangle_{\partial\Omega} d\tau$$

because the term of first order cancels

$$\left\langle\left(\frac{f_{n|\partial\Omega}}{M} - \left\langle\frac{f_n}{M}\right\rangle_{\partial\Omega}\right)h'\left(\left\langle\frac{f_n}{M}\right\rangle_{\partial\Omega}\right)\right\rangle_{\partial\Omega} = 0.$$

Then

$$\int_0^1 \left\langle \left(\frac{f_{n|\partial\Omega}}{M} - \left\langle \frac{f_n}{M} \right\rangle_{\partial\Omega} \right)^2 h'' \left(\tau \frac{f_{n|\partial\Omega}}{M} + (1-\tau) \left\langle \frac{f_n}{M} \right\rangle_{\partial\Omega} - 1 \right) \right\rangle_{\partial\Omega} d\tau$$
$$= O \left(\frac{\mathrm{Ma}_n^2 \mathrm{St}_n}{\alpha_n} \right)_{L^1(dtd\sigma_x)}$$

$$(3.9)$$

Besides we have

$$\hat{\eta}_n = \frac{2}{\mathrm{Ma}_n} \sqrt{\frac{\alpha_n}{\mathrm{St}_n}} 1_{\Sigma_+} \left(\sqrt{\frac{f_{n|\partial\Omega}}{M}} - \sqrt{\left\langle \frac{f_n}{M} \right\rangle_{\partial\Omega}} \right)$$

$$= \frac{2}{\mathrm{Ma}_n} \sqrt{\frac{\alpha_n}{\mathrm{St}_n}} 1_{\Sigma_+} \left(\frac{f_{n|\partial\Omega}}{M} - \left\langle \frac{f_n}{M} \right\rangle_{\partial\Omega} \right) \left(\sqrt{\frac{f_{n|\partial\Omega}}{M}} + \sqrt{\left\langle \frac{f_n}{M} \right\rangle_{\partial\Omega}} \right)^{-1}$$

and

$$h'' \left(\tau \frac{f_{n|\partial\Omega}}{M} + (1-\tau) \left\langle \frac{f_n}{M} \right\rangle_{\partial\Omega} - 1 \right) = \frac{1}{\tau f_{n|\partial\Omega}/M + (1-\tau)\langle f_n/M \rangle_{\partial\Omega}}.$$

Therefore,

$$\frac{4\alpha_n}{\mathrm{Ma}_n^2 \mathrm{St}_n} \left(\frac{f_{n|\partial\Omega}}{M} - \left\langle \frac{f_n}{M} \right\rangle_{\partial\Omega} \right)^2 h'' \left(\tau \frac{f_{n|\partial\Omega}}{M} + (1-\tau) \left\langle \frac{f_n}{M} \right\rangle_{\partial\Omega} - 1 \right) \geq \hat{\eta}_n^2.$$

Plugging this inequality in (3.9) leads to the expected L^2 estimate on $\hat{\eta}_n$.
□

3.1.3 The Entropy Dissipation Bound

Because of Boltzmann's H theorem, we expect the entropy dissipation bound

$$\forall t \geq 0, \quad \frac{1}{\mathrm{StKn}} \int_0^t \int D(f)(s,x)dxds \leq H_{in}$$

to control the distance between the distribution function f and the set of all
Maxwellian distributions.

In the Absence of Further Estimates

on the distribution function f, the relaxation mechanism for the (homogeneous) Boltzmann equation is however not sufficiently well understood to deduce from this dissipation bound a precise estimate on the distance between f and $\mathcal{M}_f$. This is another reason why the compressible Euler limit has (at present time) no mathematical justification in full generality.

In the Perturbative Framework,

the results by Hilbert [65] and Grad [59] on the linearized collision operator allow to control the distance between the fluctuation g defined by (3.2) (or the renormalized fluctuation $\hat{g}$ defined by (3.5)) and its projection on hydrodynamic modes. Those results are the matter of the next section.

In Both Situations,

a consequence of pointwise inequalities satisfied by the nonlinearity which defines the entropy dissipation is the following estimate on the renormalized collision kernel :

Lemma 3.1.5 *Let (f_n) be a sequence of measurable, a.e. nonnegative distribution functions such that*

$$\forall t \geq 0, \quad \int_0^t \int D(f_n)(s,x)dxds \leq C\mathrm{Ma}_n^2\mathrm{Kn}_n\mathrm{St}_n.$$

Then, the family of renormalized collision terms $(\hat{q}_n)$ defined by

$$\hat{q}_n = \frac{1}{\mathrm{Ma}_n\sqrt{\mathrm{St}_n\mathrm{Kn}_n}}\frac{1}{M}Q(\sqrt{Mf_n},\sqrt{Mf_n}) \tag{3.10}$$

is bounded in $L^2(\nu^{-1}Mdvdxdt)$, where ν is the collision frequency defined by

$$\nu(v) = \iint M_*B(v-v_*,\sigma)dv_*d\sigma.$$

Proof. By Cauchy-Schwarz' inequality, we have

$$\hat{q}^2 \leq \frac{1}{\mathrm{Ma}^2\mathrm{StKn}}\left(\iint \frac{1}{MM_*}(\sqrt{MM_*ff_*} - \sqrt{M'M'_*f'f'_*})^2B(v-v_*,\sigma)dv_*d\sigma\right)$$
$$\times \left(\iint \frac{1}{MM_*}M_*^2B(v-v_*,\sigma)dv_*d\sigma\right)$$
$$\leq \frac{1}{\mathrm{Ma}^2\mathrm{StKn}}\frac{\nu}{M}\left(\iint (\sqrt{ff_*} - \sqrt{f'f'_*})^2B(v-v_*,\sigma)dv_*d\sigma\right)$$

From the elementary inequality

$$(x-y)\log\frac{x}{y} \geq 4(\sqrt{x}-\sqrt{y})^2, \quad x,y > 0$$

we then deduce that

$$\int_0^t \iint \hat{q}^2(s,x,v)\nu^{-1}M(v)dxdvds \leq \frac{1}{\mathrm{Ma}^2\mathrm{StKn}}\int_0^t \int D(f)(s,x)dsdx \leq C,$$

which is the announced result. $\qquad\square$

3.2 Properties of the Collision Operator : Relaxation Towards Equilibrium and Regularization in v Variables

3.2.1 Some Results in the Non Perturbative Framework

Let us start by informally discussing some features which may help the trend to equilibrium, or on the contrary make it more difficult, both from the physical and from the mathematical points of view.

First of all the *distribution tails* are usually at the origin of the worst difficulties. By distribution tails, we mean how fast the distribution function decreases as $|v| \to \infty$, or $|x| \to \infty$. This is not only a technical point; Bobylev has shown that large tails could be a true obstacle to a good trend to equilibrium for the Boltzmann equation. More precisely, he proved the following result [11]. Consider the spatially homogeneous Boltzmann equation with Maxwell cross-section

$$\partial_t f = Q(f, f)$$
$$f_{|t=0} = f_{in} \tag{3.11}$$

and fix the mass, momentum and energy of the initial datum f_{in}. Let M be the corresponding equilibrium state. Then, for any $\varepsilon > 0$ one can construct an initial datum $f_{\varepsilon,in}$ such that the associated solution $f_\varepsilon \equiv f_\varepsilon(t, v)$ of the homogeneous Cauchy problem (3.11) satisfies

$$\forall t \geq 0, \quad \|f_\varepsilon(t) - M\|_{L^1(\mathbf{R}^3)} \geq K_\varepsilon e^{-\varepsilon t}, \quad K_\varepsilon > 0.$$

Note however that most of the discrepancy between f_ε and M is located at very high velocities. This illustrates the general fact that precise "experimental" information about rates of convergence to equilibrium is very difficult to have, if one wants to take into account distribution tails.

Next, it is clear that the more collisions there are, the more likely convergence is found to be fast. This is why the size of the cross-section does matter, in particular difficulties arise in the study of hard potentials because of the *vanishing of the cross-section* at zero relative velocities. However studies of the linearized operator ([65],[59]) show that in principle, one could expect an exponential decay to equilibrium for the spatially homogeneous Boltzmann equation with hard potentials (under strong control on the distribution tails).

In order to derive rigorously hydrodynamic limits of the Boltzmann equation, we need to understand the effects of the relaxation process, especially to establish quantitative variants of the mechanism of decreasing of the entropy. More precisely, one would like to prove an *entropy-entropy dissipation inequality* : this is a functional inequality of the type

$$D(f) \geq \Theta(H(f|M))$$

where $H \mapsto \Theta(H)$ is some continuous function, strictly positive when $H > 0$. Such an inequality, coupled with Boltzmann's H theorem would indeed imply that any solution to the spatially homogeneous Boltzmann equation satisfies $H(f|M)(t) \to 0$ as $t \to +\infty$, and one would be further able to compute an explicit rate of convergence.

An old conjecture of Cercignani, formulated at the beginning of the eighties, was that the spatially homogeneous Boltzmann equation would satisfy a linear entropy-entropy dissipation inequality

$$D(f) \geq 2\lambda(f)H(f|M) \qquad (3.12)$$

for some $\lambda(f)$ depending on f only via some estimates of moments, Sobolev regularity, lower bound. Bobylev and Cercignani [12] have actually disproved this conjecture, by considering distributions close to the equilibrium M, with a very tiny bump at large velocities.

However, Mouhot [87] has recently established that the exponential trend to equilibrium which should be implied by (3.12), namely

$$\|f(t) - M\|_{L^1} \leq Ce^{-\mu t} \qquad (3.13)$$

holds true in the particular case of hard spheres (and for more general hard potentials with cut-off, under a very strong decay condition). The idea of the proof is to combine linear and nonlinear techniques : quantitative estimates of exponential decay on the evolution semi-group associated to the linearized collision operator (to be detailed in the next paragraph) are used to estimate the rate of convergence when the solution is close to equilibrium (where the linear part of the collision operator is dominant), whereas the existing nonlinear entropy method, combined with some L^1 a priori estimates, is used to estimate the rate of convergence for solutions far from equilibrium.

Such a method should open new perspectives in the field of compressible hydrodynamic limits, for which the understanding of the nonlinear relaxation process is crucial. Of course, this would suppose to obtain pointwise estimates on the moments of the solution to the spatially inhomogeneous Boltzmann equation, which remains an outstanding problem.

In order to avoid this additional difficulty, in all situations requiring some control on the nonlinear relaxation process, we will consider the BGK equation instead of the Boltzmann equation, namely

$$
\begin{aligned}
& \mathrm{St}\partial_t f + v \cdot \nabla_x f = \frac{1}{\mathrm{Kn}}(\mathcal{M}_f - f), \\
& \mathcal{M}_f(t, x, v) = \frac{R(t, x)}{(2\pi T(t, x))^{3/2}} \exp\left(-\frac{|v - U(t, x)|^2}{2T(t, x)}\right), \\
& R(t, x) = \int f(t, x, v)\,dv, \quad RU(t, x) = \int vf(t, x, v)\,dv, \\
& R(|U|^2 + 3T)(t, x) = \int |v|^2 f(t, x, v)\,dv,
\end{aligned}
\qquad (3.14)
$$

which is the simpler relaxation model associated with the Boltzmann equation.

3.2.2 Coercivity of the Linearized Collision Operator

The perturbative framework is again much more propicious for a mathematical study. Considering indeed some fluctuation f around a global equilibrium state M

$$f = M(1 + \mathrm{M}ag),$$

it is easy to see that the relaxation process is governed by the linearized collision operator $\mathcal{L}_M$ defined by

$$
\begin{aligned}
\mathcal{L}_M g &\overset{\text{def}}{=} -\frac{2}{M} Q(M, Mg) \\
&= \int_{\mathbf{R}^3} dv^* \int_{S^2} d\sigma M_* B(v - v^*, \sigma)(g + g_* - g' - g'_*),
\end{aligned}
\tag{3.15}
$$

which has been extensively studied by Hilbert [65], then Grad [59] and Caflisch [23].

Because of the translation and scaling invariance of the collision kernel, we will actually restrict our attention in the sequel to the case where M is the reduced centered Gaussian

$$M(v) = \frac{1}{(2\pi)^{3/2}} \exp\left(-\frac{|v|^2}{2}\right).$$

Indeed, if τ_w and m_λ denote respectively the translation and scaling isometries on $L^1(\mathbf{R}^3)$ defined by

$$\tau_w f(v) = f(v - w), \quad (m_\lambda f)(v) = \lambda^{-3} f(\lambda^{-1} w)$$

one has

$$Q(\tau_w f, \tau_w f) = \tau_w Q(f, f), \quad Q(m_\lambda f, m_\lambda f) = \lambda m_\lambda Q(f, f).$$

We then deduce that

$$\mathcal{L}_{M_{R_0, U_0, T_0}}(\phi) = (R_0 \sqrt{T_0}) \tau_{U_0} m_{\sqrt{T_0}} \mathcal{L}_M (m_{1/\sqrt{T_0}} \tau_{-U_0} \phi). \tag{3.16}$$

Hilbert's Decomposition

In order to establish the coercivity of the linearized collision operator $\mathcal{L}_M$, the first step is to introduce *Hilbert's decomposition* [65], showing that $\mathcal{L}_M$ is just a compact perturbation of some multiplication operator :

Proposition 3.2.1 *Assume that B satisfies Grad's cut-off assumption (2.8) for some $\beta \in [0, 1]$. Then the linearized collision operator $\mathcal{L}_M$ defined by (3.15) can be decomposed as*

$$\mathcal{L}_M g(v) = \nu(|v|) g(v) - \mathcal{K} g(v)$$

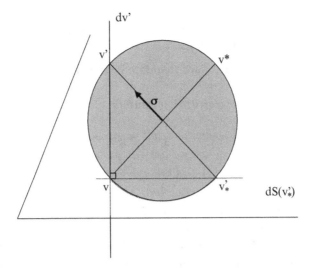

Fig. 3.1. Carleman's parametrization

where $\mathcal{K}$ is a compact integral operator on $L^2(Mdv)$ and $\nu = \nu(|v|)$ is a scalar function called the collision frequency that satisfies, for some $C > 1$,

$$0 < \nu_- \leq \nu(|v|) \leq C(1 + |v|)^\beta .$$

Sketch of proof of Proposition 3.2.1. The method is based on a clever change of variables sometimes called "Carleman's collision parametrization" although it goes back to Hilbert [65].

One changes variables in the integrals defining ν and $\mathcal{K}$, by using the transformation (see Figure 3.1)

$$(v_*, \sigma) \in \mathbf{R}^3 \times S^2 \mapsto (v', v'_*) \in C,$$

where

$$C = \{(v', v'_*) \in \mathbf{R}^3 \times \mathbf{R}^3 \ / \ (v - v'_*) \cdot (v' - v) = 0\}.$$

This transformation sends the measure $|(v - v_*) \cdot \sigma| dv_* d\sigma$ on the measure $dv' dS(v'_*)$, where dS is the surface element on the plane orthogonal to $(v' - v)$ passing through v.

With this change of variables, we obtain exact and convenient form for the function ν

$$\nu(v) = 2\pi \iint B(|v - v_*|, \cos\theta) M(v_*) \sin\theta \, d\theta \, dv_*$$

As for the integral operator $\mathcal{K}$, one first computes its integral kernel k. More precisely, one further splits the operator $\mathcal{K}$ as $\mathcal{K} = -\mathcal{K}_1 + \mathcal{K}_2$, where

$$\mathcal{K}_1 g(v) = \int_{\mathbf{R}^3} dv_* \int_{S^2} d\sigma \, M_* B(v - v_*, \sigma) g_* ,$$

and

$$K_2 g(v) = \int_{\mathbf{R}^3} dv_* \int_{S^2} d\sigma \, M_* B(v - v_*, \sigma)(g' + g'_*) \,.$$

The integral kernels k_1 and k_2 are therefore defined by

$$k_1(v, w) = 2\pi \int B(|v - w|, \cos\theta) M(w) \sin\theta d\theta,$$

$$k_2(v, w) = \frac{2}{|v - w|^2} \exp\left(-\frac{1}{4}|w|^2 + \frac{1}{4}|v|^2 - \frac{1}{8}|v - w|^2\right)$$

$$\times \int_{\eta \perp (w-v)} B\left(|v - w|^2 + |\eta|^2)^{1/2}, \left(1 + \frac{|\eta|^2}{|v - w|^2}\right)^{-\frac{1}{2}}\right) M\left(\eta + \frac{1}{2}(v + w)\right) d\eta$$

That $\mathcal{K}_1$ and $\mathcal{K}_2$ are self-adjoint is easily seen on these formulas. That $\mathcal{K}_1$ is compact on $L^2(Mdv)$ is obvious; that $\mathcal{K}_2$ is also compact on $L^2(Mdv)$ follows from observing that $\mathcal{K}_2^4$ is in the Hilbert-Schmidt class on $L^2(Mdv)$. To see this, one computes from k_2 the integral kernel of $\mathcal{K}_2^4$, say $k_2^{(4)}$, and observe that

$$(v, w) \mapsto k_2^{(4)}(v, w) \frac{M(v)^{1/2}}{M(w)^{1/2}}$$

belongs to $L^2(\mathbf{R}^3 \times \mathbf{R}^3; dvdw)$.   □

The Relative Coercivity Estimate

With the above preliminary results, we then establish the main property of the linearized collision operator $\mathcal{L}_M$, i.e. that it satisfies the Fredholm alternative in some weighted L^2 space.

Proposition 3.2.2 *Assume that B satisfies Grad's cut-off assumption (2.8) for some $\beta \in [0, 1]$. Then the linear collision operator $\mathcal{L}_M$ defined by (3.15) is a nonnegative unbounded self-adjoint operator on $L^2(Mdv)$ with domain*

$$\mathcal{D}(\mathcal{L}_M) = \{g \in L^2(Mdv) \,|\, \nu g \in L^2(Mdv)\} = L^2(\mathbf{R}^3; \nu M(v)dv)$$

and nullspace

$$\mathrm{Ker}(\mathcal{L}_M) = \mathrm{span}\{1, v_1, v_2, v_3, |v|^2\} \,.$$

Moreover the following coercivity estimate holds: there exists $C > 0$ such that, for each $g \in \mathcal{D}(\mathcal{L}_M) \cap (\mathrm{Ker}(\mathcal{L}_M))^\perp$

$$\int g\mathcal{L}_M g(v)M(v)dv \geq C\|g\|^2_{L^2(M\nu dv)} \,.$$

Sketch of Proof of Proposition 3.2.2.
   • The first step consists in characterizing the nullspace of $\mathcal{L}_M$. It must contain the *collision invariants* since the integrand in

$$\mathcal{L}_M g = \iint (g + g_* - g' - g'_*)B(v - v_*, \sigma)M_* dv_* d\sigma$$

vanishes identically if $g(v) = 1, v_1, v_2 v_3$ or $|v|^2$. Conversely, the same symmetries of the collision integral as in section 1 of Chapter 2 imply that

$$\int \psi \mathcal{L}_M g M dv =$$
$$\frac{1}{4} \iiint (\psi + \psi_* - \psi' - \psi'_*)(g + g_* - g' - g'_*)B(v - v_*, \sigma)M_* dv_* d\sigma \,.$$

Letting $g = \psi$ implies that $\mathcal{L}_M$ is a nonnegative self-adjoint operator on the weighted space $L^2(Mvdv)$. In particular, if g belongs to the nullspace of $\mathcal{L}_M$,

$$\frac{1}{4} \iiint (g + g_* - g' - g'_*)^2 B(v - v_*, \sigma)M_* dv_* dv d\sigma = 0 \,,$$

so that, for almost all $(v_*, \sigma) \in \mathbf{R}^3 \times S^2$

$$g + g_* = g' + g'_*.$$

In other words, g is a collision invariant, which entails that g is a linear combination of $1, v_1, v_2, v_3$ and $|v|^2$.

- Next we prove the *coercivity estimate*. First the multiplication operator $g \mapsto \nu g$ is self-adjoint on $L^2(\mathbf{R}^3; Mvdv)$ and has continuous spectrum which consists of the numerical range of ν, i.e. $\nu(\mathbf{R}^+) \subset [\nu_-, +\infty)$ where

$$\nu_- = \inf_{v \in \mathbf{R}^3} \nu(|v|) > 0.$$

By Weyl's theorem, as $\mathcal{K}$ is self-adjoint and compact on $L^2(\mathbf{R}^3; Mdv)$, the spectrum of $\mathcal{L}_M$ consists of $\nu(\mathbf{R}^+)$ and of a sequence of eigenvalues in the interval $[0, \nu_-]$ with ν_- as only possible accumulation point.

In particular, there exists a smallest positive element λ_1 of the spectrum of $\mathcal{L}_M$, called the *spectral gap*, and one has

$$\int g \mathcal{L}_M g(v) M(v) dv \geq \lambda_1 \|g\|^2_{L^2(Mdv)}$$

for each $g \in \mathcal{D}(\mathcal{L}_M) \cap (\mathrm{Ker}(\mathcal{L}_M))^\perp$.

The identity

$$\int g \mathcal{L}_M g M(v) dv = \int g^2 \nu M(v) dv - \int g \mathcal{K} g M(v) dv$$

together with the continuity of $\mathcal{K}$ and the coercivity estimate above imply the stronger, weighted estimate announced in the statement of Proposition 3.2.2.

$\square$

Remark 3.2.3 *Because $\mathcal{L}_M$ is a Fredholm operator, one can define $\mathcal{L}_M^{-1}$ on the orthogonal complement $(\mathrm{Ker}\mathcal{L}_M)^{\perp}$ of $\mathrm{Ker}\mathcal{L}_M$, especially on the following quantities*

$$\Phi(v) = v \otimes v - \frac{1}{3}|v|^2 \, \mathrm{Id}, \quad \Psi(v) = \frac{1}{2}v(|v|^2 - 5),$$

which are the kinetic versions of the momentum flux and heat flux, and therefore play a fundamental role in the study of hydrodynamic limits.

Note that, using some invariance properties of $\mathcal{L}_M$ (due to the isotropy of the collision process), one can obtain some additional information on the structure of

$$\tilde{\Phi} = \mathcal{L}_M^{-1}\Phi \text{ and } \tilde{\Psi} = \mathcal{L}_M^{-1}\Psi, \tag{3.17}$$

already mentioned in the works of the physicists Chapman and Cowling [33].

Desvillettes and Golse [43] have then proved that there exist two scalar functions φ and ψ such that

$$\tilde{\Phi}(v) = \varphi(|v|)\Phi(v) \text{ and } \tilde{\Psi}(v) = \psi(|v|)\Psi(v).$$

Golse and the author have further established [57] a polynomial growth estimate on both functions φ and ψ. This additional structure explains in particular why the viscosity μ and heat conductivity κ obtained in (2.22) are nonnegative scalar fields.

Control of the Relaxation in Perturbative Regimes

Consider, as in the previous section, solutions of the Boltzmann equation such that

$$H_{in} = \iint Mh \left(\mathrm{Ma} g_{in}\right)(x,v)dxdv = O(\mathrm{Ma}^2),$$

and define the corresponding renormalized fluctuations by (3.5)

$$\hat{g} = \frac{2}{\mathrm{Ma}}\left(\sqrt{\frac{f}{M}} - 1\right),$$

so that

$$\forall t \geq 0, \quad \iint M|\hat{g}|^2(t,x,v)dxdv \leq \frac{2H_{in}}{\mathrm{Ma}^2} = O(1).$$

We also recall that the fluctuation

$$g = \hat{g} + \frac{\mathrm{Ma}}{4}\hat{g}^2$$

is therefore bounded in $L_t^{\infty}(L^1(Mdvdx))$.

Then, the entropy dissipation bound, coupled with the coercivity estimate stated in the previous paragraph, provides the following control on the relaxation :

Lemma 3.2.4 *For each sequence* $\mathrm{Ma}_n \to 0$, *let* (f_n) *be a sequence of measurable, a.e. nonnegative distribution functions such that*

$$\forall t \geq 0, \quad H(f_n|M)(t) \leq C\mathrm{Ma}_n^2,$$

and

$$\forall t \geq 0, \quad \int_0^t \int D(f_n)(s,x)dxds \leq C\mathrm{Ma}_n^2 \mathrm{Kn}_n \mathrm{St}_n .$$

Then, the family of fluctuations $(\hat{g}_n)$ *defined by*

$$\hat{g}_n = \frac{2}{\mathrm{Ma}_n}\left(\sqrt{\frac{f_n}{M}} - 1 \right),$$

satisfies

$$\|\hat{g}_n - \varPi\hat{g}_n\|_{L^2(Mdv)} \leq O(\sqrt{\mathrm{Kn}_n\mathrm{St}_n})_{L^2_{t,x}} + O(\mathrm{Ma}_n)\|\hat{g}_n\|^2_{L^2(Mdv)}, \qquad (3.18)$$

where $\varPi$ *denotes the orthogonal projection on* $\mathrm{Ker}\mathcal{L}_M$.

Proof. In order to simplify the presentation, we first introduce some fictitious collision integrals $\tilde{\mathcal{L}}_M$ and $\tilde{Q}$, obtained from $\mathcal{L}_M$ and Q by replacing the original cross-section B with

$$\tilde{B}(v - v_*,\sigma) = \frac{B(v - v_*,\sigma)}{1 + \iint B(v - v_*,\sigma)M_*dv_*d\sigma}.$$

Note that the corresponding collision frequency

$$\tilde{\nu}(v) = \iint M_*\tilde{B}(v - v_*,\sigma)dv_*d\sigma$$

is therefore bounded from up and below.

Start then from the elementary formula

$$M\tilde{\mathcal{L}}_M\hat{g}_n = \frac{\mathrm{Ma}_n}{2}\tilde{Q}(M\hat{g}_n, M\hat{g}_n) - \frac{2}{\mathrm{Ma}_n}\tilde{Q}(\sqrt{Mf_n}, \sqrt{Mf_n}) \qquad (3.19)$$

Mutiplying both sides of this equation by $\hat{g}_n$ and using the coercivity estimate stated in Proposition 3.2.2 leads to

$$\begin{aligned}
&\|\hat{g}_n - \varPi\hat{g}_n\|_{L^2(Mdv)} \\
&\leq \frac{C\mathrm{Ma}_n}{2}\left\| \frac{1}{M}\tilde{Q}(M\hat{g}_n, M\hat{g}_n) \right\|_{L^2(Mdv)} + 2C\sqrt{\mathrm{St}_n\mathrm{Kn}_n}\,\|\tilde{q}_n\|_{L^2(Mdv)}
\end{aligned}$$

where $\tilde{q}_n$ denotes the renormalized collision integral

$$\tilde{q}_n = \frac{1}{\mathrm{Ma}_n\sqrt{\mathrm{St}_n\mathrm{Kn}_n}}\frac{1}{M}\tilde{Q}(\sqrt{Mf_n}, \sqrt{Mf_n}).$$

The same arguments as in the proof of Lemma 3.1.5 show that

$$\int_0^t \iint \tilde{q}_n^2(s,x,v)M(v)dxdvds \le \frac{1}{\text{Ma}_n^2\text{St}_n\text{Kn}_n} \int_0^t \int D(f_n)(s,x)dsdx.$$

By the *continuity properties of the quadratic collision operator* (see [59] for instance), we further have

$$\left\|\frac{1}{M}\tilde{Q}(M\hat{g}_n, M\hat{g}_n)\right\|_{L^2(Mdv)} \le C\|\hat{g}_n\|_{L^2(Mdv)}^2.$$

Combining both estimates leads to

$$\|\hat{g}_n - \Pi\hat{g}_n\|_{L^2(Mdv)}$$
$$\le C\sqrt{\text{St}_n\text{Kn}_n}\|\tilde{q}_n\|_{L^2(Mdv)} + C\text{Ma}_n\|\hat{g}_n\|_{L^2(Mdv)}^2$$

which is the expected inequality.   □

3.2.3 Improving Integrability with Respect to the v Variables

An important consequence of the previous control on the relaxation is to provide further integrability with respect to v-variables on the (renormalized) fluctuation $\hat{g}_n$. We have indeed the following lemma :

Lemma 3.2.5 *Let* (Ma_n), (Kn_n) *and* (St_n) *be sequences such that*

$$\text{Ma}_n \to 0, \quad \text{Kn}_n\text{St}_n \le \text{Ma}_n^2.$$

Let (f_n) *be a sequence of measurable, a.e. nonnegative distribution functions such that*
$$\forall t \ge 0, \quad H(f_n|M)(t) \le C\text{Ma}_n^2,$$

and

$$\forall t \ge 0, \quad \int_0^t \int D(f_n)(s,x)dxds \le C\text{Ma}_n^2\text{Kn}_n\text{St}_n.$$

Then, the family of fluctuations $(\hat{g}_n)$ *defined by*

$$\hat{g}_n = \frac{2}{\text{Ma}_n}\left(\sqrt{\frac{f_n}{M}} - 1\right),$$

is such that $((1+|v|)^p M|\hat{g}_n|^2)$ *is uniformly integrable in* v *on* $[0,T] \times K \times \mathbf{R}^3$ *for each* $T > 0$, *each compact* $K \subset \Omega$ *and each* $p < 2$, *meaning that*

$$\lim_{\eta \to 0}\sup_n \int_0^T \int_K \left(\sup_{|A|\le\eta}\int_A (1+|v|)^p M|\hat{g}_n|^2(t,x,v)dv\right)dtdx = 0.$$

Proof. The crucial idea behind this result is to decompose $\hat{g}_n$ according to

$$\hat{g}_n = (\hat{g}_n - \Pi\hat{g}_n) + \Pi\hat{g}_n.$$

- From the L^2 bound on $\hat{g}_n$ and the explicit formula for $\Pi\hat{g}_n$ we deduce that

$$\hat{g}_n(1 + |v|)^p\Pi\hat{g}_n = O(1)_{L_t^\infty(L_x^1(L^r(Mdv)))} \text{ for } r < 2. \qquad (3.20)$$

- It remains then to estimate

$$(1 + |v|)^p(\hat{g}_n - \Pi\hat{g}_n)\hat{g}_n$$

for $p < 2$, using the control on the relaxation provided by Lemma 3.2.4.

By Young's inequality, we have for each $\delta > 0$

$$\begin{aligned}
(1 + |v|)^p\hat{g}_n^2 &\leq \frac{\delta^2}{\mathrm{Ma}_n^2}\left|\frac{f_n}{M} - 1\right|\frac{(1 + |v|)^p}{\delta^2} \\
&\leq \frac{\delta^2}{\mathrm{Ma}_n^2}h\left(\frac{f_n}{M} - 1\right) + \frac{\delta^2}{\mathrm{Ma}_n^2}\exp\left(\frac{(1 + |v|)^p}{\delta^2}\right)
\end{aligned} \qquad (3.21)$$

so that for each $q < +\infty$

$$(1 + |v|)^{p/2}\hat{g}_n = O(\delta)_{L_t^\infty(L^2(Mdvdx))} + O\left(\frac{C_{p,q,\delta}}{\mathrm{Ma}_n}\right)_{L_{t,x}^\infty(L^q(Mdv))}$$

where $C_{p,q,\delta}$ is some nonnegative constant depending only on p, q and δ.

We therefore have

$$\begin{aligned}
(1 + |v|)^p(\hat{g}_n - \Pi\hat{g}_n)\hat{g}_n &= O(\delta)_{L_t^\infty(L^2(Mdvdx))}(1 + |v|)^{p/2}(\hat{g}_n - \Pi\hat{g}_n) \\
&+ O\left(\frac{C_{p,q,\delta}}{\mathrm{Ma}_n}\right)_{L_{t,x}^\infty(L^q(Mdv))}(1 + |v|)^{p/2}(\hat{g}_n - \Pi\hat{g}_n)
\end{aligned} \qquad (3.22)$$

from which we deduce that

$$\begin{aligned}
\int_A(1 + |v|)^p|(\hat{g}_n - \Pi\hat{g}_n)\hat{g}_n|Mdv &= O(\delta)_{L_t^\infty(L_x^2)}\left(\int_A(1 + |v|)^p|\hat{g}_n|^2Mdv\right)^{1/2} \\
&+ O(\delta)_{L_t^\infty(L_x^2)}\left(\int_A(1 + |v|)^p|\Pi\hat{g}_n|^2Mdv\right)^{1/2} \\
+ \frac{C_{p,q,\delta}}{\mathrm{Ma}_n}\left(O(\sqrt{\mathrm{Kn}_n\mathrm{St}_n})_{L_{t,x}^2} + O(\mathrm{Ma}_n)\|\hat{g}_n\|_{L^2(Mdv)}^2\right)&\left(\int_A M(1 + |v|)^{p\tilde{q}/2}dv\right)^{1/\tilde{q}}
\end{aligned} \qquad (3.23)$$

with $1/q + 1/\tilde{q} = 1/2$.

- Plugging (3.20) and (3.23) together, we get

$$\begin{aligned}
\int_A(1 + |v|)^p&|\hat{g}_n|^2Mdv \\
&= \left(\int_A Mdv\right)^{1-\frac{1}{r}}O(1)_{L_t^\infty(L_x^2)} + O(\delta)_{L_t^\infty(L_x^2)}\left(\int_A(1 + |v|)^p|\hat{g}_n|^2Mdv\right)^{1/2} \\
&+ \left(\int_A M(1 + |v|)^{p\tilde{q}/2}dv\right)^{1/\tilde{q}}\frac{C_{p,q,\delta}}{\mathrm{Ma}_n}\left(O(\sqrt{\mathrm{Kn}_n\mathrm{St}_n})_{L_{t,x}^2} + O(\mathrm{Ma}_n)_{L_t^\infty(L_x^1)}\right)
\end{aligned}$$

Choosing δ small enough, and using Cauchy-Schwarz' inequality to master the second term in the right-hand side, we obtain the expected local equiintegrability with respect to v.   □

Remark 3.2.6 *In the general non perturbative framework, the entropy dissipation bound is also expected to give some further integrability with respect to v variables on the distribution f_n.*

Of course it is not clear that the second term in the decomposition

$$f_n = \mathcal{M}_{f_n} + (f_n - \mathcal{M}_{f_n})$$

can be controlled by the entropy dissipation. However, instead of using the local Maxwellian $\mathcal{M}_{f_n}$ in the above decomposition, one can consider the local pseudo-equilibrium

$$A_{f_n} = \frac{1}{\rho_n}\tilde{Q}^+(f_n, f_n)$$

where $\rho_n = \int f_n dv$ and $\tilde{Q}^+$ is the gain part of some fictitious collision operator $\tilde{Q}$.

For a suitable choice of $\tilde{Q}$ one expects $A_{f_n} - f_n$ to be controlled by the entropy dissipation and thus to be small in the fast relaxation limit, whereas A_{f_n} should have further integrability in the v variables due to the compactness properties of $\tilde{Q}^+$ (shown by Lions [72] or Bouchut and Desvillettes [15]).

3.3 Properties of the Free Transport Operator : Dispersion and Averaging Lemmas

According to our principle that the structure of the limiting hydrodynamic system should be recognized in the corresponding scaled Boltzmann equation, we expect further regularity estimates to hold, at least in the hydrodynamic regime leading to the incompressible Navier-Stokes system. In such a viscous regime, the entropy dissipation bound gives actually a control on the collision term, which can be considered as a source term for the free transport equation. It remains then to understand the smoothing process leading to these regularity estimates.

The fundamental operator in kinetic theory is the advection or free-transport operator

$$\mathrm{St}\partial_t + v \cdot \nabla_x$$

which is the prototype of hyperbolic operators. It expresses that a particle of velocity v is advected of a distance $dx = vdt$ during the infinitesimal time interval dt. The solution of the advection equation

$$\mathrm{St}\partial_t f + v \cdot \nabla_x f = 0$$

is actually given by the method of characteristics

$$f(t, X(t, x, v), V(t, x, v)) = f_{in}(x, v)$$

or equivalently

$$f(t, x, v) = f_{in}(X(-t, x, v), V(-t, x, v))$$

where

$$\frac{dX}{dt} = V, \quad \frac{dV}{dt} = 0, \quad (X, V)(0, x, v) = (x, v)$$

(possibly supplemented with boundary conditions). A consequence of that formula is that singularities of the initial or boundary data are propagated at finite speed $|v|/\text{St}$ inside the domain where the equation is posed. In particular, one cannot hope to gain regularity or compactness on the distribution function f itself.

With a view towards hydrodynamic limits we then anticipate that only macroscopic variables, i.e. moments in v of the distribution function of the type

$$\int_{\mathbf{R}^3} f(t, x, v)\varphi(v)dv$$

for some test function φ, should be regular or compact. This is precisely the essence of the class of results known as "*velocity averaging*", to be described below.

Let us first concentrate on the case of the whole space $x \in \mathbf{R}^3$.

3.3.1 L^2 Averaging Lemmas

The first regularity results bearing on moments of the solution of a transport equation were obtained in the L^2 setting. Indeed, the key idea of the proof of such results is a kind of reduction to the one dimensional case, which is especially simple when expressed in terms of Fourier variables.

The regularity of the averages is actually due to the ellipticity of the symbol of the free transport operator $\text{St}\partial_t + v \cdot \nabla_x$ outside from a small subset of the velocity space $\mathbf{R}_v^3$. In Fourier variables, one has indeed

$$i(\text{St}\tau + v \cdot \xi)\mathcal{F}f(\tau, \xi, v) = \mathcal{F}S,$$

where S denotes some source term : this seems to indicate that $\mathcal{F}f$ should have good decay properties with respect to the ξ variables.

If $|\text{St}\tau + v \cdot \xi| > \alpha$, one obtains some regularity on the corresponding part of the solution (all the more that α is large). If $|\text{St}\tau + v \cdot \xi| \leq \alpha$, the contribution of the corresponding part of the solution to the average is small (all the smaller that α is small). See Figure 3.2.

Translating the previous argument into a precise mathematical statement, and choosing α to optimize the resulting estimate leads to the following by now classical proposition (obtained independently by Golse, Lions, Perthame, Sentis [53] and Agoshkov [1]).

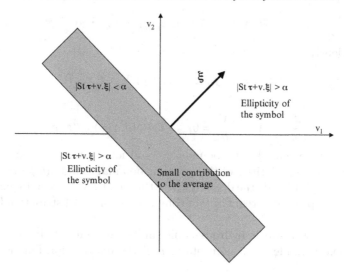

Fig. 3.2. Symbol of the free transport operator

Proposition 3.3.1 *Let* $f \in L^2(\mathbf{R}_t \times \mathbf{R}_x^3 \times \mathbf{R}_v^3)$ *be the solution of the free-transport equation*

$$\mathrm{St}\partial_t f + v \cdot \nabla_x f = S.$$

Then, for all compactly supported test function $\varphi \in L^\infty(\mathbf{R}_v^3)$, *the following regularity estimate holds*

$$\left\| \int f\varphi(v)dv \right\|_{L^2(\mathbf{R}, H^{1/2}(\mathbf{R}^3))} \leq C\|f\|_{L^2(\mathbf{R}_t \times \mathbf{R}_x^3 \times \mathbf{R}_v^3)}^{1/2} \|S\|_{L^2(\mathbf{R}_t \times \mathbf{R}_x^3 \times \mathbf{R}_v^3)}^{1/2}$$

for some nonnegative constant C *depending only on* φ.

Proof. Let us introduce as previously the time and space Fourier variables (τ, ξ). The Fourier transform of the moment is therefore

$$\int \mathcal{F}f(\tau, \xi, v)\varphi(v)dv.$$

As explained above, the estimate is obtained by splitting this integral into two contributions

$$\left| \int \mathcal{F}f(\tau, \xi, v)\varphi(v)dv \right| \leq \int \mathbf{1}_{|\mathrm{St}\tau + v\cdot\xi| \leq \alpha} |\mathcal{F}f(\tau, \xi, v)\varphi(v)|dv$$

$$+ \int \mathbf{1}_{|\mathrm{St}\tau + v\cdot\xi| > \alpha} |\mathcal{F}f(\tau, \xi, v)\varphi(v)|dv.$$

By Cauchy-Schwarz' inequality, we then get

$$\left| \int \mathcal{F}f(\tau, \xi, v)\varphi(v) dv \right| \le \left(\int |\mathcal{F}f(\tau, \xi, v)|^2 dv \right)^{\frac{1}{2}} \left(\int 1_{|\mathrm{St}\tau + v \cdot \xi| \le \alpha} |\varphi(v)|^2 dv \right)^{\frac{1}{2}}$$

$$+ \left(\int |\mathcal{F}S(\tau, \xi, v)|^2 dv \right)^{\frac{1}{2}} \left(\int \frac{1_{|\mathrm{St}\tau + v \cdot \xi| > \alpha}}{|\mathrm{St}\tau + v \cdot \xi|^2} |\varphi(v)|^2 dv \right)^{\frac{1}{2}}$$

$$\le C \|\mathcal{F}f\|_{L^2(\mathbf{R}_v^3)} \left(\frac{\alpha}{|\xi|} \right)^{1/2} + C \|\mathcal{F}S\|_{L^2(\mathbf{R}_v^3)} \left(\frac{1}{\alpha |\xi|} \right)^{1/2}$$

where C is some nonnegative constant depending only on $\|\varphi\|_\infty$ and of the support of φ. Thus, using Plancherel's identity and choosing the truncation parameter $\alpha = \|\mathcal{F}S\|_{L^2} / \|\mathcal{F}f\|_{L^2}$ leads to the optimal estimate.   □

Remark 3.3.2 *This fundamental result can be extended in several directions :*
- *a refined estimate shows that on can also gain some regularity with respect to the scaled time variable t/St;*
- *the same compactness result still holds for generalized advection operators*

$$\mathrm{St}\partial_t + a(v) \cdot \nabla_x$$

(up to some loss in the regularity exponent), provided that they satisfy a non concentration condition of the following type

$$\forall R > 0, \quad \lim_{\varepsilon \to 0} \sup_{e \in S^2} \left| \{v \in \mathbf{R}^3 \, / \, |v| \le R \text{ and } |a(v) \cdot e| \le \varepsilon \} \right| = 0.$$

In particular this condition is satisfied for the relativistic free transport operator $a(v) = v / \sqrt{1 + |v|^2}$.
- *more generally, if f and S belong to $L^p(\mathbf{R} \times \mathbf{R}^3 \times \mathbf{R}^3)$, one has the similar regularity estimate*

$$\left\| \int f\varphi(v) dv \right\|_{L^p(\mathbf{R}, W^{s,p}(\mathbf{R}^3))} \le C(\|f\|_{L^p(\mathbf{R}_t \times \mathbf{R}_x^3 \times \mathbf{R}_v^3)} + \|S\|_{L^p(\mathbf{R}_t \times \mathbf{R}_x^3 \times \mathbf{R}_v^3)})$$

where $s = \min(1/p, 1/p')$ (see [53] for the detailed interpolation argument). In particular, if $p = 1$ or $+\infty$, there is no general velocity averaging result.

That velocity averaging fails in L^1 and L^∞ is not due to some technical deficiency in the proof, but rather to a real difficulty linked with concentration phenomena, as shown by the following counterexample taken from [53].

Let $S_n \equiv S_n(t, x, v)$ be a bounded sequence in $L^1(\mathbf{R} \times \mathbf{R}^3 \times \mathbf{R}^3)$ such that

$$S_n \to \mathrm{St}\varphi'(t)\delta_{x - \mathrm{St}^{-1} v_0 t} \otimes \delta_{v - v_0}$$

for some test function $\varphi \in C_c^\infty(\mathbf{R})$, where $v_0 \ne 0$. Let $f_n \equiv f_n(t, x, v)$ be the sequence of solutions of the transport equation

$$\mathrm{St}\partial_t f_n + v \cdot \nabla_x f_n = S_n, \quad (t, x, v) \in \mathbf{R} \times \mathbf{R}^3 \times \mathbf{R}^3 .$$

The method of characteristics provides

$$f_n(t, x, v) = \mathrm{St}^{-1} \int_{-\infty}^{t} S_n(s, x - \mathrm{St}^{-1}v(t - s), v)ds$$

and thus, for each $\phi \in C_c(\mathbf{R}^3)$

$$\int_{\mathbf{R}^3} \phi(x) \left(\int_{\mathbf{R}^3} f_n(t, x, v)dv \right) dx$$

$$= \mathrm{St}^{-1} \int_{-\infty}^{t} \iint_{\mathbf{R}^3 \times \mathbf{R}^3} S_n(s, z, v)\phi(z + \mathrm{St}^{-1}v(t - s))dvdzds$$
$$\to \varphi(t)\phi(\mathrm{St}^{-1}v_0 t)$$

as $n \to +\infty$. Hence

$$\int_{\mathbf{R}^3} f_n dv \text{ converges to some density carried by the half plane } \mathbf{R} \times \mathbf{R}_+ v_0$$

in the weak sense of measures as $n \to +\infty$. In particular

$$\int_{\mathbf{R}^3} f_n dv \text{ is not relatively compact in } L^1_{loc}(\mathbf{R} \times \mathbf{R}^3)$$

although

$$\|f_n\|_{L^1(\mathbf{R} \times \mathbf{R}^3 \times \mathbf{R}^3)} \le C,$$
$$\|\mathrm{St}\partial_t f_n + v \cdot \nabla_x f_n\|_{L^1(\mathbf{R} \times \mathbf{R}^3 \times \mathbf{R}^3)} \le C.$$

This counterexample suggests in particular that one should try by all means to control concentration effects.

3.3.2 Dispersive Properties of the Free-Transport Operator

A first remark is that it is actually sufficient to control the concentration effects in the v variables, the non concentration in the x variables arising then as a consequence of some *dispersion properties* of the free transport operator.

These dispersion properties are linked with the propagation of the volumes in the phase space. Indeed, by the free transport, the domains of the phase space become stretched in the x direction, according to the scheme represented in Figure 3.3.

A set of "small measure in x" is changed into a set of "small measure in v", which can be referred to as a *mixing property* of the free transport operator.

More precisely, using the exact representation of the solution to the free transport equation, and an appropriate change of variables, Castella and Perthame have obtained the following gain of integrability with respect to x variables [26] :

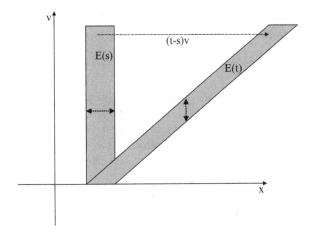

Fig. 3.3. Dispersion by the free transport operator

Proposition 3.3.3 *Let* χ *be the solution of the free-transport equation*

$$\partial_s \chi + v \cdot \nabla_x \chi = 0 \ on \ \mathbf{R} \times \mathbf{R}_x^3 \times \mathbf{R}_v^3,$$
$$\chi_{|s=0} = \chi_0 \ on \ \mathbf{R}_x^3 \times \mathbf{R}_v^3.$$

Then, for all pair $(p, q) \in [1, +\infty]$ *with* $p \leq q$, *the following pointwise estimate holds*

$$\forall t \in \mathbf{R}^*, \quad \|\chi(s)\|_{L_x^q(L_v^p)} \leq |s|^{-3\left(\frac{1}{p} - \frac{1}{q}\right)} \|\chi_0\|_{L_x^p(L_v^q)}.$$

Proof. In the absence of external force and source term, the solution χ is given by

$$\chi(s, x, v) = \chi_0(x - sv, v),$$

which immediately leads to

$$\|\chi(s)\|_{L_x^\infty(L_v^1)} \leq |s|^{-3} \|\chi_0\|_{L_x^1(L_v^\infty)},$$

and more generally to

$$\|\chi(s)\|_{L_x^\infty(L_v^p)} \leq |s|^{-3/p} \|\chi_0\|_{L_x^p(L_v^\infty)}.$$

On the other hand, the conservation of the L^p norm gives

$$\|\chi(s)\|_{L_{x,v}^p} = \|\chi_0\|_{L_{x,v}^p}.$$

Thus, by interpolation,

$$\|\chi(s)\|_{L_x^q(L_v^p)} \leq |s|^{-\frac{3}{p}\left(1 - \frac{p}{q}\right)} \|\chi_0\|_{L_x^p(L_v^q)},$$

which is the expected estimate. □

Remark 3.3.4 • *The extension of such a result to the relativistic case is not so trivial : due to the Jacobian of the change of variables*

$$v \mapsto \frac{v}{\sqrt{1+v^2}}$$

the resulting dispersion estimate will involve moments of the initial data, which has however no consequence in the context of hydrodynamic limits.

• *This result will be indeed used to transfer some local equiintegrability from v variables (coming from the control on the relaxation) to x variables, meaning that in all proofs we will restrict our attention to a bounded set of velocities.*

3.3.3 L^1 Averaging Lemmas

Combining both techniques, we are then able to improve the velocity averaging results, extending them in the L^1 setting under an additional condition of equiintegrability with respect to v, to be recalled now.

We say that a bounded sequence (ψ_n) of $L_{x,v}^1$ is *uniformly equiintegrable with respect to v* if

$$\lim_{\eta \to 0} \sup_n \int_{\mathbf{R}^3} \left(\sup_{|A| \leq \eta} \int_A |\psi_n(x,v)| dv \right) dx = 0.$$

As a consequence of Proposition 3.3.3, we indeed establish, using Dunford-Pettis' criterion, that the uniform integrability in v is a L^1 weak compactness criterion for the solutions to the free transport equation.

Proposition 3.3.5 *Let (f_n) be a bounded sequence in $L^\infty(\mathbf{R}^+, L^1(\mathbf{R}^3 \times \mathbf{R}^3))$ such that*

$$(\text{St}\partial_t f_n + v \cdot \nabla_x f_n) \text{ is bounded in } L^1(\mathbf{R}^+ \times \mathbf{R}^3 \times \mathbf{R}^3),$$
$$(f_n) \text{ is uniformly equiintegrable in } v.$$

Then the sequence (f_n) is uniformly equiintegrable (in all variables), and thus weakly compact by Dunford-Pettis' criterion.

Proof. This result was stated and proved in [56], extending an earlier remark by the author [92]. It relies on Proposition 3.3.3, coupled with Green's formula.

Without loss of generality, we assume that f_n is nonnegative, and that all the f_n's are supported in the same compact K of $\mathbf{R} \times \mathbf{R}^3 \times \mathbf{R}^3$.

• The first step consists in splitting any "small" measurable subset A of K in three subsets A_0, A_1 and A_2 which are "small" respectively in the t, x and v variables.

In order to do that, we define

$$A_0 = \{(t,x,v) \in A \ / \iint 1_A(t,x,v)dxdv > |A|^{1/2}\},$$

$$A_1 = \{(t, x, v) \in A \setminus A_0 \ / \int 1_A(t, x, v) dv > |A|^{1/4}\},$$

and

$$A_2 = A \setminus (A_0 \cup A_1).$$

From the *Bienaymé-Tchebichev inequality*

$$|\{x / |\chi(x)| \geq y\}| \leq \frac{1}{y} \int |\chi(x)| dx, \qquad (3.24)$$

we therefore deduce the following estimates

$$\begin{aligned}
\|1_{A_0}\|_{L_t^1(L_{x,v}^\infty)} &\leq |A|^{1/2}, \\
\|1_{A_1}\|_{L_t^\infty(L_x^1(L_v^\infty))} &\leq |A|^{1/4}, \\
\|1_{A_2}\|_{L_{t,x}^\infty(L_v^1)} &\leq |A|^{1/4}.
\end{aligned} \qquad (3.25)$$

In particular, from the uniform equiintegrability in the t and v variables, we get that the first and third terms in the decomposition

$$\begin{aligned}
\iiint_A f_n(t, x, v) dt dx dv &= \iiint_{A_0} f_n(t, x, v) dt dx dv \\
&+ \iiint_{A_1} f_n(t, x, v) dt dx dv \\
&+ \iiint_{A_2} f_n(t, x, v) dt dx dv
\end{aligned}$$

go to zero as $|A| \to 0$. It remains then to transfer some of the equiintegrability in the v variables on the x variables, to establish that the second term goes also to zero as $|A| \to 0$.

• The second step relies then crucially on the mixing properties of the free transport operator.

We start by introducing an additional parameter s_*, to be optimized in the sequel. We therefore consider the solution $\chi \equiv \chi(s, t, x, v)$ of the free transport equation

$$\begin{aligned}
\partial_s \chi + St \partial_t \chi + v \cdot \nabla_x \chi = 0, \quad s > 0, \ (t, x, v) \in \mathbf{R} \times \mathbf{R}^3 \times \mathbf{R}^3, \\
\chi(0, t, x, v) = 1_{A_1}(t, x, v), \quad (t, x, v) \in \mathbf{R} \times \mathbf{R}^3 \times \mathbf{R}^3.
\end{aligned}$$

Clearly, $\chi(s, t, x, v) = 1_{A_1}(t - Sts, x - sv, v)$. A slight adaptation of Proposition 3.3.3 and the second estimate in (3.25) imply that

$$\|\chi(s_*, ., ., .)\|_{L_{t,x}^\infty(L_v^1)} \leq \frac{1}{|s_*|^3} \|1_{A_1}\|_{L_t^\infty(L_x^1(L_v^\infty))} \leq \frac{|A|^{1/4}}{|s_*|^3} \qquad (3.26)$$

Applying Green's formula to the integral

$$\int_0^{s_*} ds \iint_K dt dx dv f_n(t, x, v)(\partial_s + St \partial_t + v \cdot \nabla_x) \chi(s, t, x, v)$$

leads then to

$$\iiint_{A_1} f_n(t,x,v)\,dt\,dx\,dv = \iiint_K f_n\chi(s_*,t,x,v)\,dt\,dv\,dx$$
$$- \int_0^{s_*} \iiint_K \chi(s,t,x,v)(\mathrm{St}\partial_t + v\cdot\nabla_x)f_n(t,x,v)\,ds\,dt\,dx\,dv$$

(3.27)

From the uniform L^1 bound on $(\mathrm{St}\partial_t f_n + v\cdot\nabla_x f_n)$ we deduce that the second term in the right side of (3.27) tends to zero as $|s_*| \to 0$. Furthermore, using the equiintegrability of (f_n) with respect to v and the estimate (3.26) shows that the first term tends to zero as $|A|^{1/4}/|s_*|^3 \to 0$. Therefore, choosing for instance $|s^*| = |A|^{1/24}$, we get that

$$\iiint_{A_1} f_n(t,x,v)\,dt\,dx\,dv \to 0 \text{ as } |A| \to 0,$$

which, coupled with the previous results, leads to the expected equiintegrability. □

Equipped with this preliminary result, we can now state the main L^1 velocity averaging result, to be used in the context of the incompressible Navier-Stokes limit of the Boltzmann equation.

Theorem 3.3.6 *Let (f_n) be a bounded sequence in $L^\infty(\mathbf{R}^+, L^1(\mathbf{R}^3 \times \mathbf{R}^3))$ such that*

$$(\mathrm{St}\partial_t f_n + v\cdot\nabla_x f_n) \text{ is bounded in } L^1(\mathbf{R}^+ \times \mathbf{R}^3 \times \mathbf{R}^3),$$
$$(f_n) \text{ is uniformly equiintegrable in } v.$$

Then, for all compactly supported test function $\varphi \in L^\infty(\mathbf{R}_v^3)$, the sequence $(\int f_n\varphi(v)\,dv)$ is strongly compact with respect to the x variables, in the sense that, for all compact subset $\tilde{K}$ of $\mathbf{R}^+ \times \mathbf{R}^3$,

$$\lim_{|\eta|\to 0} \iint_{\tilde{K}} \left| \int f_n(t,x,v)\varphi(v)\,dv - \int f_n(t,x+\eta,v)\varphi(v)\,dv \right| dx\,dt = 0.$$

Proof. Without loss of generality, we again assume that f_n is nonnegative.
• The first step consists in establishing the compactness statement assuming that both families (f_n) and $(\mathrm{St}\partial_t f_n + v\cdot\nabla_x f_n)$ are weakly compact in L^1, which has been done in the original paper by Golse, Lions, Perthame and Sentis [53].
Define

$$\phi_n = f_n + (\mathrm{St}\partial_t + v\cdot\nabla_x)f_n$$

and for all $\lambda > 0$

$$\phi_n^{\lambda,+} = \phi_n \mathbf{1}_{\phi_n > \lambda}, \quad \phi_n^{\lambda,-} = \phi_n \mathbf{1}_{\phi_n \le \lambda}.$$

We also denote by $f_n^{\lambda,+}$ and $f_n^{\lambda,-}$ the solutions to

$$\phi_n^{\lambda,+} = f_n^{\lambda,+} + (\mathrm{St}\partial_t + v \cdot \nabla_x)f_n^{\lambda,+}, \quad \phi_n^{\lambda,-} = f_n^{\lambda,-} + (\mathrm{St}\partial_t + v \cdot \nabla_x)f_n^{\lambda,-}.$$

From the explicit formula of the resolvent $R = (\mathrm{Id} +\mathrm{St}\partial_t + v \cdot \nabla_x)^{-1}$

$$R\phi(t,x,v) = \int_0^{+\infty} e^{-\sigma}\phi(t - \mathrm{St}\sigma, x - \sigma v, v)d\sigma$$

we deduce that

$$(f_n^{\lambda,-}) \text{ and } ((\mathrm{St}\partial_t + v \cdot \nabla_x)f_n^{\lambda,-}) \text{ are uniformly bounded in } L_{t,x,v}^2$$

by the L^2-norm of $\phi_n^{\lambda,-}$, i.e. by a constant depending only on λ. Applying the L^2 averaging lemma leads then to

$$\left\| \int f_n^{\lambda,-}\varphi(v)dv \right\|_{L_t^2(H_x^{1/2})} \leq C_\lambda.$$

On the other hand, the same explicit formula for the resolvent R shows that

$$\|f_n^{\lambda,+}\|_{L_{t,x,v}^1} \leq \|\phi_n^{\lambda,+}\|_{L_{t,x,v}^1}$$

and thus converges to 0 as $\lambda \to +\infty$ by the equiintegrability assumption. In particular, for all compact $K \subset [0,T] \times \mathbf{R}^3$ and for all $\delta > 0$, there exists λ such that

$$\iint_K \left| \int f_n^{\lambda,+}(t,x,v)\varphi(v)dv \right| dxdt \leq \delta.$$

Then, choosing η sufficiently small,

$$\iint_{\tilde{K}} \left| \int f_n^{\lambda,-}(t,x,v)\varphi(v)dv - \int f_n^{\lambda,-}(t,x+\eta,v)\varphi(v)dv \right| dxdt \leq \delta.$$

Combining both estimates leads to the expected regularity result.

• The second step consists in extending this result in the case when $(\mathrm{St}\partial_t f_n + v \cdot \nabla_x f_n)$ is only assumed to be bounded in L^1.

To do that, we introduce another truncation parameter, as follows. For each $\mu > 0$, set

$$R_\mu = (\mu\,\mathrm{Id} +\mathrm{St}\partial_t + v \cdot \nabla_x)^{-1}.$$

Using the explicit formula

$$R_\mu\phi(t,x,v) = \int_0^{+\infty} e^{-\mu\sigma}\phi(t - \mathrm{St}\sigma, x - \sigma v, v)d\sigma,$$

one easily checks that, for each $p \in [1, +\infty]$,

$$\|R_\mu\phi\|_{L_{t,x,v}^p} \leq \frac{\|\phi\|_{L_{t,x,v}^p}}{\mu}.$$

Write

$$f_n = R_\mu(\mu f) + R_\mu((St\partial_t + v \cdot \nabla_x)f_n) = f_n^{\mu,1} + f_n^{\mu,2}$$

so that

$$\int f_n \varphi(v) dv = \int f_n^{\mu,1} \varphi(v) dv + \int f_n^{\mu,2} \varphi(v) dv,$$

with

$$f_n^{\mu,1} = \mu R_\mu f_n \text{ and } f_n^{\mu,2} = R_\mu((St\partial_t + v \cdot \nabla_x)f_n).$$

Since $(St\partial_t + v \cdot \nabla_x)f_n$ is uniformly bounded in $L^1_{t,x,v}$, the second term on the right hand side of the equality above can be made arbitrarily small in $L^1_{t,x,v}$ for some $\mu > 0$ large enough : for all $\delta > 0$, there exists μ such that

$$\iint_K \left| \int f_n^{\mu,2}(t,x,v)\varphi(v) dv \right| dx dt \leq \delta.$$

For such a μ, the first term on the right hand side of the equality above satisfies :

$$\iint_{\tilde K} \left| \int f_n^{\mu,1}(t,x,v)\varphi(v) dv - \int f_n^{\mu,1}(t,x+\eta,v)\varphi(v) dv \right| dx dt \leq \delta$$

provided that η is sufficiently small (depending on δ and μ). Combining both estimates leads to the expected regularity result.

Note that these two first steps rely actually on the following simple characterization of relatively compact sets :

$$\mathcal{K} \text{ relatively compact} \iff \forall \varepsilon > 0, \quad \exists \mathcal{K}_\varepsilon \text{ compact}, \, d(\mathcal{K}, \mathcal{K}_\varepsilon) \leq \varepsilon. \quad (3.28)$$

• The final step consists then in relaxing the equiintegrability assumption on (f_n), assuming only that (f_n) is uniformly equiintegrable in the (t,v) variables. This follows directly from Proposition 3.3.5. □

3.3.4 The Case of a Spatial Domain with Boundaries

In the case of a spatial domain with boundaries, the boundary conditions do not allow in general to get an explicit formula for the characteristics, and therefore for the solution of the free transport equation.

Nevertheless, a simple localization argument allows to extend all the previous results, starting for instance from the a priori estimates obtained in Appendix B.

The fundamental averaging result states :

Proposition 3.3.7 *Let Ω be some smooth open domain of $\mathbf{R}^3$. Let $f \in L^2_{loc}(\mathbf{R}_t \times \Omega_x \times \mathbf{R}^3_v)$ be the solution to the free-transport equation*

$$\mathrm{St}\partial_t f + v \cdot \nabla_x f = S,$$

for some $S \in L^2_{loc}(\mathbf{R}_t \times \Omega_x \times \mathbf{R}^3_v)$. Then, for all compactly supported test function $\varphi \in L^\infty(\mathbf{R}^3_v)$, all $T > 0$ and all compact subset K_x of Ω, the following regularity estimate holds

$$\left\| \int f\varphi(v)dv \right\|_{L^2([-T,T],H^{1/2}(K_x))} \le C$$

for some nonnegative constant C depending only on φ, T and K_x.

Sketch of proof of Proposition 3.3.7. We will not give extensive proof of that result, insofar as it is essentially based on the same arguments as in the case of the whole space. Let us just explain how we introduce the truncation, and what are the main subsequent modifications.

Let Ω be some smooth open domain. Then, for any compact $K \subset \mathbf{R} \times \Omega$, there exist some compact $\tilde{K} \subset \mathbf{R} \times \Omega$ and some C^∞ function χ_1 (see Figure 3.4) such that

$$\chi_1 \equiv 1 \text{ on } K, \text{ and } \chi_1 \equiv 0 \text{ outside from } \tilde{K}.$$

In the same way, we can truncate large velocities $|v| \ge R$ (far from the support of φ) by some smooth function $\chi_2 \in C^\infty_c(\mathbf{R}^3_v)$. We then denote $\chi = \chi_1\chi_2$.

For any $f \in L^2_{loc}(\mathbf{R}^+ \times \Omega \times \mathbf{R}^3_v)$, we define $f\chi$ on $\mathbf{R}^+ \times \Omega \times \mathbf{R}^3_v$, then extend it to $\mathbf{R}^+ \times \Omega \times \mathbf{R}^3_v$ by 0. We then have

$$\|f\chi\|_{L^2(\mathbf{R}^+ \times \mathbf{R}^3 \times \mathbf{R}^3)} \le C\|f\|_{L^2(\tilde{K} \times B_R)}.$$

In the same way,

$$\mathrm{St}\partial_t(f\chi) + v \cdot \nabla_x(f\chi) = S\chi + f(\mathrm{St}\partial_t\chi + v \cdot \nabla_x\chi),$$

so that

$$\|(\mathrm{St}\partial_t + v \cdot \nabla_x)(f\chi)\|_{L^2(\mathbf{R}^+ \times \mathbf{R}^3 \times \mathbf{R}^3)}$$
$$\le C\left(\|S\|_{L^2(\tilde{K} \times B_R)} + \|f\|_{L^2(\tilde{K} \times B_R)} \right)$$

We then apply Proposition 3.3.1, from which we deduce that the moment

$$\int f\chi\varphi dv = \chi_1 \int f\varphi dv$$

belongs to $L^2(\mathbf{R}^+, H^{1/2}(\mathbf{R}^3))$. We obviously have the same regularity on the subdomain K on which $f\chi_1 = f$. □

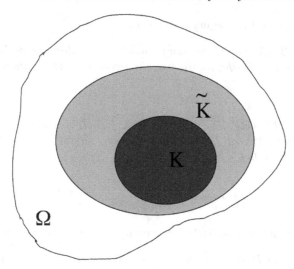

Fig. 3.4. The truncation function χ_1

The same procedure allows actually to extend any averaging result on smooth domains, even in the L^1 case. Note indeed that the so-called "dispersion property" used to obtain the L^1 result is not really a global result in the sense that it does not use the decay estimate as time tends to infinity, but rather a short time estimate. Combined with Gronwall's lemma, it gives in fact a (local) *mixing property* with respect to x and v. That is why it can be extended to bounded domains :

Theorem 3.3.8 *Let Ω be some smooth open domain. Let (f_n) be a bounded sequence in $L^\infty_{loc}(\mathbf{R}^+, L^1_{loc}(\Omega \times \mathbf{R}^3))$ such that*

$(\mathrm{St}\partial_t f_n + v \cdot \nabla_x f_n)$ *is bounded in* $L^1_{loc}(\mathbf{R}^+ \times \Omega \times \mathbf{R}^3)$,
(f_n) *is locally uniformly equiintegrable in v.*

Then, for all compactly supported test function $\varphi \in L^\infty(\mathbf{R}^3_v)$, the sequence $(\int f_n \varphi(v)dv)$ is strongly compact with respect to x variables, in the sense that, for all compact subset K of $\mathbf{R}^+ \times \Omega$,

$$\lim_{|\eta| \to 0} \iint_K \left| \int f_n(t,x,v)\varphi(v)dv - \int f_n(t,x+\eta,v)\varphi(v)dv \right| dxdt = 0.$$

4

The Incompressible Navier-Stokes Limit

At the present time, the incompressible Navier-Stokes limit is the only hydrodynamic asymptotics of the Boltzmann equation for which an optimal convergence result is known (and for which we are actually able to implement all the mathematical tools presented in the previous chapter). By "optimal", we mean here that this convergence result

- holds globally in time;
- does not require any assumption on the initial velocity profile;
- does not assume any constraint on the initial thermodynamic fields;
- takes into account boundary conditions, and describes their limiting form.

4.1 Convergence Result : From the Boltzmann Equation to the Incompressible Navier-Stokes-Fourier System

4.1.1 Mathematical Theories for the Incompressible Navier-Stokes Equations

Before giving the precise mathematical statement of this convergence result, let us first recall some basic facts about the homogeneous incompressible Navier-Stokes equations.

The first equation expresses the incompressibility constraint

$$\nabla \cdot u = 0,$$

and the second equation is the local conservations of momentum

$$\partial_t u + (u \cdot \nabla_x)u + \nabla p = \mu \Delta u,$$

where the macroscopic density R and the viscosity μ of the gas are assumed to be constant, and the pressure p is the Lagrange multiplier associated with the incompressibility constraint. The diffusion operator is expected to have a smoothing effect, which is linked to the fact that informations propagate with

L. Saint-Raymond, *Hydrodynamic Limits of the Boltzmann Equation*,
Lecture Notes in Mathematics 1971, DOI: 10.1007/978-3-540-92847-8_4,
© Springer-Verlag Berlin Heidelberg 2009

an infinite speed, or in other words to the fact that the bulk velocity of the gas is negligible compared to the speed of sound.

The fundamental estimates for the study of hydrodynamic models are those coming from physics, namely the *conservations of total mass and energy*, as well as the possible *decrease of entropy*. In the case of homogeneous incompressible fluids, these estimates provide bounds of L^2 type on the velocity field. They are formally obtained by a simple computation of hilbertian analysis, remarking that the scalar product of a gradient field by a divergence-free vector field (with zero mass flux at the boundary) is zero :

Proposition 4.1.1 *Let $u \equiv u(t,x)$ be a solution of the incompressible Navier-Stokes equations*

$$\partial_t u + (u \cdot \nabla_x)u + \nabla p = \mu \Delta u, \quad \nabla_x \cdot u = 0, \tag{4.1}$$

$$u_{|t=0} = u_{in}, \tag{4.2}$$

that is sufficiently smooth (for instance in $C(\mathbf{R}^+, H^1(\Omega))$) and satisfies the zero mass flux condition at the boundary

$$n \cdot u_{|\partial\Omega} = 0,$$

where n is the outward unit normal to $\partial\Omega$. Then the following energy estimate holds :

$$\|u(t)\|_{L^2(\Omega)}^2 + 2\mu \int_0^t \|\nabla u(s)\|_{L^2(\Omega)}^2 ds + 2 \int_0^t \int_{\partial\Omega} \Sigma : n \otimes u(s,x) d\sigma_x ds \tag{4.3}$$
$$= \|u_{in}\|_{L^2(\Omega)}^2$$

where Σ is the stress tensor defined by

$$\Sigma = \mu(\nabla u + (\nabla u)^T) - p \, \mathrm{Id},$$

and σ_x the surface measure on $\partial\Omega$.

Another important property which is useful for the study of this type of models is the *scaling invariance*, namely the fact that for any solution u to (4.1), and any $\lambda > 0$, the function u_λ defined by

$$u_\lambda(t,x) = \lambda u(\lambda^2 t, \lambda x)$$

is still a solution to (4.1). Although this property has no physical meaning, it allows on the one hand to determine critical functional spaces, in which one can expect to establish global existence of solutions with small data (see for instance the work by Cannone, Meyer and Planchon [25]), and on the other hand to study qualitative properties of the solutions (for instance their large time behaviour, using self-similar solutions to (4.1) [49]).

Weak Solutions "à la Leray"

By using only the L^2 energy estimate, Leray [70] has proved the global existence of weak solutions to (4.1)(4.2). Indeed the additional regularity coming
from the dissipation term in the energy inequality allows to take weak limits
in some approximation scheme.

This argument is reminiscent from the gain of spatial regularity on the
moments of the Boltzmann equation, obtained from the entropy dissipation
bound by the averaging lemma : the presence at some position $x \in \Omega$ of
particles with different velocities leads to some *friction*, and consequently to
some correlation between the bulk velocities at adjoining positions. We will see
later in this chapter a more precise mathematical formulation of this analogy.

Theorem 4.1.2 *Let* $u_{in} \in L^2(\Omega)$ *be a divergence free vector field. Then there
exists (at least) one global weak solution* $u \in L^2_{loc}(\mathbf{R}^+, H^1(\Omega)) \cap C(\mathbf{R}^+, w -
L^2(\Omega))$ *to the incompressible Navier-Stokes equations (4.1)(4.2) supplemented
by boundary conditions, either the Dirichlet boundary condition*

$$u_{|\partial\Omega} = 0, \tag{4.4}$$

or the Navier boundary condition

$$n \cdot u_{|\partial\Omega} = 0 \text{ and } n \wedge (\Sigma \cdot n - \lambda u)_{|\partial\Omega} = 0 \tag{4.5}$$

It further satisfies the energy inequality

$$\|u(t)\|^2_{L^2(\Omega)} + 2\mu \int_0^t \|\nabla u(s)\|^2_{L^2(\Omega)} ds \leq \|u_{in}\|^2_{L^2(\Omega)} \tag{4.6}$$

in the Dirichlet case, and

$$\|u(t)\|^2_{L^2(\Omega)} + 2\mu \int_0^t \|\nabla u(s)\|^2_{L^2(\Omega)} ds + 2\lambda \int_0^t \int_{\partial\Omega} |u|^2(s, x) d\sigma_x ds \leq \|u_{in}\|^2_{L^2(\Omega)} \tag{4.7}$$

in the Navier case.

As mentioned in the previous paragraph, the Leray solutions to the incompressible Navier-Stokes equations are built by taking weak limits in some convenient approximation scheme, for instance the so-called Friedrichs scheme.

- Approximate equations are obtained by projecting (4.1) on suitable
Hilbert spaces of finite dimension, generated by eigenmodes of the Stokes
operator (with convenient boundary conditions). These ordinary differential
equations can be solved by using the Cauchy-Lipschitz theory, which further
provides some regularity estimates.

- From the uniform energy bound, on can then obtain spatial regularity
estimates on the approximate solutions, and thereby some control on their

time derivatives. By interpolation, one thus gets some strong compactness on the sequence of approximate solutions.

- Taking limits in the weak formulation of (4.1)(4.2), and in the energy inequality provides then the expected result.

Remark 4.1.3 *Note that such a method does not allow in general to obtain further regularity or stability results.*

- *In two space dimensions, it happens however that the functional spaces defined by the energy bounds, namely $L_t^\infty(L_x^2)$ and $L_t^2(\dot{H}_x^1)$, are scaling invariant. One can then prove, using methods of harmonic analysis (to be sketched in the next paragraph) that the solution is unique, and depends Lipschitz continuously on the initial data.*

- *In three space dimensions, functional spaces which are scaling invariant involve more regularity or more integrability (for instance $L_t^\infty(\dot{H}_x^{1/2})$ and $L_t^2(\dot{H}_x^{3/2})$), and thus do not correspond to (global) physical a priori estimates. Refined results due to Leray [70] and to Caffarelli, Kohn and Nirenberg [22] allow to improve regularity estimates for weak solutions, namely to prove that the Leray solutions to (4.1)(4.2) are smooth outside from small time sets (of zero Hausdorff measure), but this is not enough to obtain stability and uniqueness.*

Smooth Solutions

In 3D, considering the scaling invariant functional space $L_t^\infty(\dot{H}_x^{1/2}) \cap L_t^2(\dot{H}_x^{3/2})$, Fujita and Kato [48] have established the global existence and uniqueness of solutions under a suitable smallness assumption on the initial data, and the local existence and uniqueness of solutions for general initial data. Their result has then be extended to more general scaling invariant functional spaces ([25],[27]).

For the sake of simplicity, we restrict here our attention to the case of the whole space, which avoids to determine the suitable formulation of boundary conditions (compatible with the required regularity).

Theorem 4.1.4 *Let $u_{in} \in H^{1/2}(\mathbf{R}^3)$ be a divergence free vector field. Then there exists a unique maximal solution*

$$u \in C([0, t^*), H^{1/2}(\mathbf{R}^3)) \text{ with } \nabla u \in L_{loc}^2([0, t^*), H^{1/2}(\mathbf{R}^3))$$

to the incompressible Navier-Stokes equations (4.1)(4.2).
Furthermore, if $t^* < +\infty$,

$$\int_0^{t^*} \|\nabla u(t)\|_{\dot{H}^{1/2}(\mathbf{R}^3)}^2 dt = +\infty.$$

The key argument in the proof is the stability estimate established by Serrin [99], stating that all weak solutions to (4.1)(4.2) must coincide with the strong solution if the latter does exist.

- The first step consists actually in establishing the global existence and uniqueness of the solution for any initial data in a subset of the critical space (here a small ball of $\dot{H}^{1/2}(\mathbf{R}^3)$). For such solutions, the precised energy estimate allows to obtain a global a priori bound in the critical space. This regularity estimate, coupled with the stability argument, implies the uniqueness.

- One can then prove the local existence and uniqueness of the solution for any initial data in the critical space, by splitting this solution as the sum of a solution of the corresponding Stokes equation, and a small remainder in $\dot{H}^{1/2}(\mathbf{R}^3)$ (which can be dealt with as in the first step).

Strong-Weak Stability

The uniqueness of the solution is therefore essentially based on a stability result, which is obtained as an energy estimate

Proposition 4.1.5 *Let u and $\tilde{u}$ be two Leray solutions to the incompressible Navier-Stokes equations (4.1) with respective initial data $u_{in}, \tilde{u}_{in} \in L^2(\mathbf{R}^3)$. Assume that, for some nonnegative $t^* > 0$, $\tilde{u}$ belongs to $L^\infty([0, t^*], H^{1/2}(\mathbf{R}^3)) \cap L^2([0, t^*], H^{3/2}(\mathbf{R}^3))$. Then, for all $t \leq t^*$,*

$$\|u(t) - \tilde{u}(t)\|_{L^2(\mathbf{R}^3)}^2 \leq \|u_{in} - \tilde{u}_{in}\|_{L^2(\mathbf{R}^3)}^2 \exp\left(\frac{C}{\mu}\int_0^t \|\nabla\tilde{u}(s)\|_{\dot{H}^{1/2}(\mathbf{R}^3)}^2 ds\right), \quad (4.8)$$

for some nonnegative constant C.

In particular, $u = \tilde{u}$ on $[0, t^*] \times \mathbf{R}^3$ if $u_{in} = \tilde{u}_{in}$.

Proof. In order to obtain the stability estimate, the first step consists in writing the equation satisfied by $w = u - \tilde{u}$:

$$\partial_t w + u \cdot \nabla w - \mu\Delta w = -\nabla p - w \cdot \nabla\tilde{u},$$

from which we deduce the formal L^2 estimate :

$$\|w(t)\|_{L^2(\mathbf{R}^3)}^2 - \|w_{in}\|_{L^2(\mathbf{R}^3)}^2$$
$$\leq -2\int_0^t \int (u \cdot \nabla)w \cdot w(s, x)dxds + 2\mu\int \Delta w \cdot w(s, x)dxds$$
$$+ 2\int_0^t \int (w \cdot \nabla)\tilde{u} \cdot w(s, x)dxds$$
$$\leq -2\mu\int_0^t \|\nabla w(s)\|_{L^2(\mathbf{R}^3)}^2 ds + 2\int_0^t \int (w \cdot \nabla)\tilde{u} \cdot w(s, x)dxds$$

because u and w are divergence free vector fields. (In order to justify this formal computation one should of course proceed by approximation, namely establish similar inequalities for the smooth approximate solutions to (4.1)(4.2), then take limits using the regularity of $\tilde{u}$ and the uniform a priori bounds coming from the energy estimate).

By Hölder's inequality, we have

$$\left| \int (w \cdot \nabla)\tilde{u} \cdot w(s,x)dx \right| \leq \|w(s)\|_{L^2(\mathbf{R}^3)} \|\nabla \tilde{u}(s)\|_{L^3(\mathbf{R}^3)} \|w(s)\|_{L^6(\mathbf{R}^3)},$$

and thus, by Sobolev's embeddings,

$$\left| \int (w \cdot \nabla)\tilde{u} \cdot w(s,x)dx \right| \leq C\|w(s)\|_{L^2(\mathbf{R}^3)} \|\nabla \tilde{u}(s)\|_{\dot{H}^{1/2}(\mathbf{R}^3)} \|\nabla w(s)\|_{L^2(\mathbf{R}^3)}.$$

From Cauchy-Schwarz' inequality we then deduce that

$$\|w(t)\|^2_{L^2(\mathbf{R}^3)} - \|w_{in}\|^2_{L^2(\mathbf{R}^3)} + 2\mu \int_0^t \|\nabla w(s)\|^2_{L^2(\mathbf{R}^3)}ds$$
$$\leq \mu \int_0^t \|\nabla w(s)\|^2_{L^2(\mathbf{R}^3)}ds + \frac{C^2}{\mu} \int_0^t \|w(s)\|^2_{L^2(\mathbf{R}^3)}\|\nabla \tilde{u}(s)\|^2_{\dot{H}^{1/2}(\mathbf{R}^3)}ds.$$

We conclude by Gronwall's lemma. □

4.1.2 Analogies with the Scaled Boltzmann Equation

• The Leray energy inequality (which allows to define global weak solutions to the incompressible Navier-Stokes equation (4.1)(4.2))

$$\|u(t)\|^2_{L^2(\mathbf{R}^3)} + 2\mu \int_0^t \|\nabla u(s)\|^2_{L^2(\mathbf{R}^3)}ds \leq \|u_{in}\|^2_{L^2(\mathbf{R}^3)}$$

and the DiPerna-Lions entropy inequality (which allows to define global renormalized solutions to the Boltzmann equation (2.18))

$$\frac{1}{\mathrm{Ma}^2} H(f|M)(t) + \frac{1}{\mathrm{KnStMa}^2} \int_0^t \int D(f)(s,x)dxds \leq \frac{1}{\mathrm{Ma}^2} H(f_{in}|M)$$

are actually very similar objects.

More precisely, it was proved by Bardos, Golse and Levermore in [5] that the Leray energy inequality is the limiting form of the DiPerna-Lions entropy inequality in incompressible viscous hydrodynamic regime

$$\mathrm{Kn} = \mathrm{Ma} = \mathrm{St} = \varepsilon \to 0,$$

since any limiting point g of the sequence of renormalized fluctuations $\hat{g}_\varepsilon$ (defined by (3.5)) satisfies

$$\frac{1}{2} \iint M|g|^2(t,x,v)dxdv \leq \liminf_{\varepsilon \to 0} \frac{1}{\varepsilon^2} H(f_\varepsilon|M)(t),$$

and

$$\int_0^t \iint M\nu^{-1}|v \cdot \nabla_x g|^2(s,x,v)dsdxdv \leq \liminf_{\varepsilon \to 0} \frac{1}{\varepsilon^4} \int_0^t \int D(f_\varepsilon)(s,x)dsdx.$$

The second bound, coupled with velocity averaging lemma established in the previous chapter, provides spatial regularity on the moments of g, as does the dissipation term in the Leray energy inequality.

This confirms the view expressed by Lions (see [74]) : "[...] the global existence result of [renormalized] solutions [...] can be seen as the analogue for Boltzmann's equation to the pioneering work on the Navier-Stokes equations by J. Leray".

Note that this analogy can be extended when considering domains with boundary (see [82]). The dissipation due to the interaction with the boundary in the case of a slipping condition arises indeed as the limiting form of the boundary term in the entropy inequality :

$$\lambda \iiint_{\Sigma_+} M(v \cdot n)(g - \langle g \rangle_{\partial\Omega})^2(s,x,v)dvd\sigma_x ds$$

$$\leq \liminf_{\varepsilon \to 0} \frac{1}{\sqrt{2\pi}\varepsilon^3} \int_0^t \int_{\partial\Omega} \left\langle h\left(\frac{f_{\varepsilon|\Sigma_+}}{M}\right) - h\left(\frac{f_{\varepsilon|\Sigma_-}}{M}\right) \right\rangle_{\partial\Omega} d\sigma_x ds$$

where $\langle g \rangle_{\partial\Omega}$ is defined as the average

$$\langle g \rangle_{\partial\Omega}(s,x) = \sqrt{2\pi} \int M(v \cdot n)_+ g(s,x,v)dv.$$

Note that the Darrozès-Guiraud information is asymptotically equal to one half of the boundary term, so that we only have

$$\frac{1}{2}\lambda \iiint_{\Sigma_+} M(v \cdot n)(g - \langle g \rangle_{\partial\Omega})^2(s,x,v)dvd\sigma_x ds$$

$$\leq \liminf_{\varepsilon \to 0} \frac{\alpha_\varepsilon}{\varepsilon^3} \int_0^t \int_{\partial\Omega} E(f_\varepsilon|M)(s,x)d\sigma_x ds$$

• Similarities also appear in the framework of smooth solutions, which is the fundamental idea in the pioneering work by Bardos and Ukai [7] on the Navier-Stokes limit of the Boltzmann equation. They have indeed proved that, for small enough initial data, the scaled Boltzmann equation (2.18) with

$$\text{Kn} = \text{Ma} = \text{St} = \varepsilon$$

has global classical solutions in the weighted Sobolev space $H_{l,k}$ (for $l > 3/2$, $k > 5/2$) defined by

$$H_{l,k} = \{f = M(1 + \varepsilon g) \,/\, \sup_{v \in \mathbf{R}^3}(1 + |v|^k)\|M^{1/2}g\|_{H^l(\mathbf{R}_x^3)} < +\infty.\}$$

Their derivation of the Navier-Stokes limit relies then on a rigorous proof of the relation between such a well-posedness result for the Boltzmann equation and the Fujita-Kato theory of the Navier-Stokes equation. The point to be stressed is that exactly the same type of assumptions are made on the initial data.

Nevertheless, as mentioned in the introduction, this result requires sharp estimates on the linearized collision operator $\mathcal{L}_M$, and overall regularity and smallness conditions on the initial data (which have no physical meaning). We have therefore decided not to explain it in details in this survey.

It would be however possible to consider the hybrid case, namely the asymptotics of renormalized solutions to the Boltzmann equation leading to some smooth solution to the incompressible Navier-Stokes equation, which is the counterpart at kinetic level of the strong-weak uniqueness principle stated in Proposition 4.1.5.

In [52], Golse, Levermore and the author have actually established a stability inequality of the following type

$$
\frac{1}{\varepsilon^2} H(f_\varepsilon | \mathcal{M}_{1,\varepsilon u,1})(t) + \frac{1}{\varepsilon^4} \int_0^t \int D(f_\varepsilon)(s,x) ds dx
$$
$$
\leq \frac{C}{\varepsilon^2} H(f_{\varepsilon,in} | \mathcal{M}_{1,\varepsilon u_{in},1}) \exp\left(C \int_0^t \|\nabla u(s)\|_{L^\infty \cap L^1(\mathbf{R}^3)} ds \right).
$$

as long as u is sufficiently smooth. This means in particular that one can expect to establish a strong convergence result for well-prepared initial data, provided that the limiting system has a (unique) strong solution. The *relative entropy method* leading to such a result will be detailed in the next chapter, in the framework of the inviscid incompressible limit of the Boltzmann equation, for which the weak compactness method fails.

Note that the strong convergence result, stating that the relative entropy

$$
\frac{1}{\varepsilon^2} H(f_\varepsilon | \mathcal{M}_{1,\varepsilon u,1})(t) \to 0,
$$

holds actually under the only assumption that the limiting Navier-Stokes system has a unique solution, satisfying the energy equality (see [5] for instance).

4.1.3 Statement of the Result

The *weak compactness method* initiated by Bardos, Golse and Levermore in [5] consists in getting the asymptotic hydrodynamic equations by taking limits in the local conservation laws associated with the Boltzmann equation, as proposed by Grad in the framework of the compressible Euler limit. In such an approach we are therefore interested only in the macroscopic parameters, and will neglect completely all problems due to initial or boundary layers, where the gas could be far from local thermodynamic equilibrium.

Theorem 4.1.6 *Let $(f_{\varepsilon,in})$ be a family of initial fluctuations around a global equilibrium M, i.e. such that*

$$\frac{1}{\varepsilon^2} H(f_{\varepsilon,in}|M) \leq C_{in}, \tag{4.9}$$

and satisfying further the weak convergences

$$\frac{1}{\varepsilon} P \int f_{\varepsilon,in} v \, dv \rightharpoonup u_{in}, \quad \frac{1}{\varepsilon} \int (f_{\varepsilon,in} - M)(\frac{1}{5}|v|^2 - 1) dv \rightharpoonup \theta_{in},$$

in $L^1_{loc}(\Omega)$, where P denotes the Leray projection onto divergence free vector fields.

Let (f_ε) be a family of renormalized solutions to the scaled Boltzmann equation

$$\varepsilon \partial_t f_\varepsilon + v \cdot \nabla_x f_\varepsilon = \frac{1}{\varepsilon} Q(f_\varepsilon, f_\varepsilon) \text{ on } \mathbf{R}^+ \times \Omega \times \mathbf{R}^3,$$
$$f_\varepsilon(0, x, v) = f_{\varepsilon,in}(x, v) \text{ on } \Omega \times \mathbf{R}^3,$$
$$f_\varepsilon(t, x, v) = (1 - \alpha_\varepsilon) f_\varepsilon(t, x, R_x v) + \sqrt{2\pi} \alpha_\varepsilon M(v) \int (v' \cdot n)_+ f_\varepsilon(t, x, v') dv' \text{ on } \Sigma_- \tag{4.10}$$

where Q is the collision operator defined by (2.7) associated with some collision kernel B satisfying Grad's cut off assumption (2.8), and where the accommodation coefficient α_ε is assumed to satisfy the scaling assumption

$$\frac{\alpha_\varepsilon}{\sqrt{2\pi}\,\varepsilon} \to \lambda \in [0, +\infty].$$

Then the family of fluctuations (g_ε) defined by $f_\varepsilon = M(1 + \varepsilon g_\varepsilon)$ is relatively weakly compact in $L^1_{loc}(dtdx, L^1((1 + |v|^2)Mdv))$, and any limit point g of (g_ε) is an infinitesimal Maxwellian

$$g = \Pi g = \rho + u \cdot v + \theta \frac{|v|^2 - 3}{2}$$

where Π is the projection on the kernel of the linearized collision operator $\mathcal{L}_M$.

Furthermore the moments of the limiting fluctuation satisfy the Navier-Stokes Fourier system, namely

$$\partial_t u + u \cdot \nabla_x u + \nabla p - \mu \Delta_x u = 0, \quad \nabla_x \cdot u = 0 \text{ on } \mathbf{R}^+ \times \Omega,$$
$$\partial_t \theta + u \cdot \nabla_x \theta - \kappa \Delta_x \theta = 0, \quad \nabla_x(\rho + \theta) = 0 \text{ on } \mathbf{R}^+ \times \Omega, \tag{4.11}$$
$$u_{|t=0} = u_{in}, \quad \theta_{|t=0} = \theta_{in} \text{ on } \Omega$$

supplemented either by the Navier boundary condition if $\lambda < +\infty$

$$(\mu(\nabla u + (\nabla u)^T) \cdot n - \lambda u) \wedge n = 0, \quad u \cdot n = 0 \text{ on } \mathbf{R}^+ \times \partial\Omega,$$
$$\kappa \frac{\partial \theta}{\partial n} - \frac{4}{5}\lambda \theta = 0 \text{ on } \mathbf{R}^+ \times \partial\Omega, \tag{4.12}$$

or by the Dirichlet boundary condition if $\lambda = +\infty$

$$u = 0, \quad \theta = 0 \text{ on } \mathbf{R}^+ \times \partial\Omega. \tag{4.13}$$

Remark 4.1.7 (i) *The solution u to the Navier-Stokes equations obtained at the limit is not necessarily a Leray solution to (4.1)(4.2) since it is not known to satisfy the energy inequality in the form (4.6) or (4.7).*

(ii) Once a weak solution

$$u \in L^2_{loc}(\mathbf{R}^+, H^1(\Omega)) \cap C(\mathbf{R}^+, w - L^2(\Omega))$$

of the incompressible Navier-Stokes equations (4.1)(4.2) is known, it is easy to see that the Fourier equation

$$\partial_t \theta + u \cdot \nabla\theta - \kappa\Delta\theta = 0$$

supplemented with some initial data $\theta_{in} \in L^2(\Omega)$ and convenient boundary conditions, has a unique weak solution

$$\theta \in L^2_{loc}(\mathbf{R}^+, H^1(\Omega)) \cap C(\mathbf{R}^+, w - L^2(\Omega))$$

satisfying further the L^2 estimate

$$\|\theta(t)\|^2_{L^2(\Omega)} + 2\kappa \int_0^t \|\nabla\theta(s)\|^2_{L^2(\Omega)} ds \leq \|\theta_{in}\|^2_{L^2(\Omega)}.$$

Note that the quantities involved here do not clearly come from physics, in particular they do not correspond to the internal part of the global energy. They actually arise when linearizing the entropy functional for fluctuations around a global equilibrium.

As mentioned in the previous paragraph, this weak convergence result can be strengthened in a strong convergence result, if there is no acoustic waves and provided that the limiting system has a unique solution satisfying the energy estimate.

In order to state this convergence result, we need the notion of *entropic convergence* introduced by Bardos, Golse and Levermore in [5].

Definition 4.1.8 *The family of fluctuations (g_ε) is said to converge entropically of order δ_ε to $g \in L^2(dxMdv)$ if and only if*

$$g_\varepsilon \rightharpoonup g \text{ weakly in } L^1_{loc}(dxMdv),$$
$$\frac{1}{\delta_\varepsilon^2} H\big(M(1 + \delta_\varepsilon g_\varepsilon)|M\big) \to \frac{1}{2}\iint Mg^2(x,v)dxdv. \qquad (4.14)$$

With this definition, we have the following result :

Theorem 4.1.9 *Let $(f_{\varepsilon,in})$ be a family of initial fluctuations around a global equilibrium M, i.e satisfying (4.9). Assume that $(g_{\varepsilon,in})$ converges entropically (of order ε) to the infinitesimal Maxwellian g_{in} defined by*

$$g_{in}(x,v) = u_{in}(x) \cdot v + \theta_{in}(x)\frac{|v|^2 - 5}{2}$$

for some given $(u_{in}, \theta_{in}) \in L^2(\Omega)$ such that

$$\nabla \cdot u_{in} = 0 \text{ in } \Omega, \quad n \cdot u_{in} = 0 \text{ on } \partial\Omega.$$

Let (f_ε) be a family of renormalized solutions to the scaled Boltzmann equation (4.10) where Q is the collision operator defined by (2.7) associated with some collision kernel B satisfying Grad's cut off assumption (2.8), and the accommodation coefficient α_ε is assumed to satisfy the scaling assumption

$$\frac{\alpha_\varepsilon}{\sqrt{2\pi}\varepsilon} \to \lambda \in [0, +\infty].$$

Assume that the Navier-Stokes Fourier system (4.11) supplemented either by the Navier condition (4.12) if $\lambda < +\infty$, or by the Dirichlet condition (4.13) in the opposite case, admits a (unique) weak solution (u, θ) satisfying the energy equality.

Then, for almost all $t > 0$, the family of fluctuations $(g_\varepsilon(t))$ defined by $f_\varepsilon = M(1 + \varepsilon g_\varepsilon)$ converges entropically to the infinitesimal Maxwellian $g(t)$ given by

$$g(t, x, v) = u(t, x) \cdot v + \theta(t, x)\frac{|v|^2 - 5}{2}.$$

4.2 The Moment Method

4.2.1 Description of the Strategy

Our goal here is to establish the convergence of appropriately scaled families of solutions to the Boltzmann equation towards solutions of the incompressible Navier-Stokes (Fourier) equations, without restrictions on the size, regularity or well-preparedness of the initial data. This means in particular that, given the state of the art about the Boltzmann equation, we consider renormalized solutions to (4.10). Note however that the method we present here could apply without important simplifications to any solution of the Boltzmann equation, since the uniform estimates come mainly from the entropy inequality.

The first results in this framework are due to Bardos, Golse and Levermore for the time independent case [5], and to Lions and Masmoudi for the time dependent problem [76]. The principle of the derivation is as follows :

- In terms of g_ε, the Boltzmann equation (4.10) becomes

$$\varepsilon\partial_t g_\varepsilon + v \cdot \nabla_x g_\varepsilon + \frac{1}{\varepsilon}\mathcal{L}_M g_\varepsilon = \frac{1}{M}Q(Mg_\varepsilon, Mg_\varepsilon), \quad (4.15)$$

where $\mathcal{L}_M$ denotes as previously the linearization of Boltzmann's collision operator at the Maxwellian state M. Therefore, multiplying (4.15) by ε and letting $\varepsilon \to 0$ suggests that (g_ε) converges in the sense of distributions to some infinitesimal Maxwellian g (see Proposition 3.2.2 in the previous chapter)

$$g(t, x, v) = \rho(t, x) + u(t, x) \cdot v + \frac{1}{2}\theta(t, x)(|v|^2 - 3).$$

- Passing to the limit in the local conservations of mass and momentum leads then to the constraints

$$\nabla_x \cdot \int Mgv dv = 0 \text{ and } \nabla_x \int Mg|v|^2 dv = 0,$$

or equivalently

$$\nabla_x \cdot u = 0 \text{ and } \nabla_x(\rho + \theta) = 0,$$

known as the *incompressibility and Boussinesq relations*.
- Recast finally the formal momentum and energy equations as

$$\partial_t \int Mg_\varepsilon v dv + \nabla_x \cdot \frac{1}{\varepsilon} \int Mg_\varepsilon \Phi(v) dv + \nabla_x \left(\frac{1}{3\varepsilon} \int Mg_\varepsilon |v|^2 dv\right) = 0,$$

$$\partial_t \int Mg_\varepsilon \frac{1}{2}(|v|^2 - 5) dv + \nabla_x \cdot \frac{1}{\varepsilon} \int Mg_\varepsilon \Psi(v) dv = 0$$

where Φ and Ψ are the momentum flux tensor and heat flux function defined by (3.17). By Remark 3.2.3 in the previous chapter and (4.15), one thus has

$$\partial_t P \int Mg_\varepsilon v dv + P\nabla_x \cdot \int (Q(Mg_\varepsilon, Mg_\varepsilon) - v \cdot \nabla_x Mg_\varepsilon) \, \tilde{\Phi}(v) dv = O(\varepsilon),$$

$$\partial_t \int Mg_\varepsilon \frac{|v|^2 - 5}{2} dv + \nabla_x \cdot \int (Q(Mg_\varepsilon, Mg_\varepsilon) - v \cdot \nabla_x Mg_\varepsilon) \, \tilde{\Psi}(v) dv = O(\varepsilon)$$

so that, using the relaxation $g_\varepsilon \sim \Pi g_\varepsilon$ where Π denotes as previously the L^2 orthogonal projection onto $\text{Ker}(\mathcal{L}_M)$, and the identity

$$Q(M\Pi g, M\Pi g) = \frac{1}{2}M\mathcal{L}_M(\Pi g)^2, \tag{4.16}$$

we can identify in both equations the convection and diffusion terms, and get in the limit $\varepsilon \to 0$ the motion and heat equations in (4.11).

In [5] or [76], such a formal process was justified assuming :
(i) the local conservations of momentum and energy, which are not guaranteed for the renormalized solutions of the Boltzmann equation;
(ii) some nonlinear estimate, namely

$$(1 + |v|^2)\frac{g_\varepsilon^2}{1 + \frac{\varepsilon}{2}g_\varepsilon} \text{ relatively weakly compact in } L^1_{loc}(dtdx, L^1(Mdv))$$

which provides both a control on large velocities v, and some equiintegrability with respect to space variables x.

In [51], Golse and Levermore have developed (in the framework of the Stokes asymptotics) an argument based on the study of the conservation defects, in order to establish that the local conservation laws hold asymptotically and to remove completely assumption (i). Such a method can be extended to the Navier-Stokes asymptotics under assumption (ii) for instance.

Hence, verifying (ii) remained the main obstruction to derive the Navier-Stokes (Fourier) equations from the Boltzmann equation. In the framework of the BGK equation (which is the relaxation model associated to the Boltzmann equation), a weak variant of (ii), which is actually sufficient to obtain a rigorous derivation of the Navier-Stokes limit, has been established by the author in [92]. It uses in a crucial way the dissipation control given by the H Theorem, namely the fact that

$$f_\varepsilon - \mathcal{M}_{f_\varepsilon} = O(\varepsilon^2),$$

combined with dispersive properties of the free-transport operator leading to L^1 velocity averaging (see Proposition 3.3.5 in the previous chapter). This argument has then been extended to the case of the Boltzmann equation for Maxwell molecules (corresponding to some bounded cross-section B) by Golse and the author [54], considering some pseudo-equilibrium A_{f_ε}, defined in terms of the gain part of the collision operator, instead of the local Maxwellian $\mathcal{M}_{f_\varepsilon}$

$$f_\varepsilon - A_{f_\varepsilon} = O(\varepsilon^2).$$

The proof can actually be simplified and generalized to all hard potentials with cut off (see [55]), by using the renormalized fluctuation

$$\hat{g}_\varepsilon \in L^\infty(\mathbf{R}^+, L^2(dxMdv))$$

defined by (3.5), and the hilbertian structure of the domain of $\mathcal{L}_M$, which allow in particular to establish in a simple way

$$\hat{g}_\varepsilon - \Pi\hat{g}_\varepsilon = O(\varepsilon),$$

(see Proposition 3.2.4 in the previous chapter).

The suitable functional framework to study the Navier-Stokes Fourier asymptotics is therefore that defined in the previous chapter, starting from the scaled entropy inequality, and using both the properties of the linearized collision operator $\mathcal{L}_M$ and the properties of the free transport operator $(\varepsilon\partial_t + v \cdot \nabla_x)$. Let us then recall the notations and basic estimates obtained for the scaled Boltzmann equation (4.10).

Two important quantities are the renormalized fluctuation $\hat{g}_\varepsilon$ and the renormalized collision integral $\hat{q}_\varepsilon$ defined by

$$\hat{g}_\varepsilon = \frac{2}{\varepsilon}\left(\sqrt{\frac{f_\varepsilon}{M}} - 1\right), \quad \hat{q}_\varepsilon = \frac{1}{\varepsilon^2}\frac{1}{M}Q(\sqrt{Mf_\varepsilon}, \sqrt{Mf_\varepsilon}) \qquad (4.17)$$

for which the relative entropy and the entropy dissipation give the following L^2 bounds

$$\|\hat{g}_\varepsilon\|^2_{L^\infty(\mathbf{R}^+, L^2(Mdvdx))} \le 2C_{in}, \quad \|\hat{q}_\varepsilon\|^2_{L^2(dtdx\nu^{-1}Mdv)} \le C_{in}. \tag{4.18}$$

By Lemmas 3.2.4 and 3.2.5 we further have the following relaxation bound and integrability estimate with respect to v

$$\|\hat{g}_\varepsilon - \Pi\hat{g}_\varepsilon\|_{L^2(Mdv)} = O(\varepsilon)_{L^2_{t,x}} + O(\varepsilon)\|\hat{g}_\varepsilon\|^2_{L^2(Mdv)}, \tag{4.19}$$

and

$$(1 + |v|^p)M|\hat{g}_\varepsilon|^2 \text{ is uniformly integrable in } v \text{ on } [0, T] \times K \times \mathbf{R}^3, \tag{4.20}$$

for all $p < 2$, all $T > 0$ and all compact $K \subset \Omega$.

From these estimates, we can easily deduce that any limiting fluctuation $g \in L^\infty(\mathbf{R}^+, L^2(dxMdv))$ is an infinitesimal Maxwellian, i.e.

$$g(t, x, v) = \Pi g(t, x, v) = \rho(t, x) + u(t, x) \cdot v + \theta(t, x)\frac{1}{2}(|v|^2 - 3),$$

and that its moments ρ, u and θ satisfy the incompressibility and Boussinesq constraints. In the sequel, we will therefore focus on the derivation of the motion and temperature equations.

4.2.2 Convergence of the Conservation Defects

Start from the scaled Boltzmann equation (4.10) renormalized relatively to M with the nonlinearity $\Gamma(z) = (z - 1)\gamma(z)$ where γ is a smooth truncation such that

$$\gamma \in C^\infty(\mathbf{R}^+, [0, 1]), \quad \gamma_{|[0,3/2]} \equiv 1, \quad \gamma(z) \le \frac{C}{1 + z} \tag{4.21}$$

With the notations γ_ε for $\gamma(f_\varepsilon/M)$ and $\hat{\gamma}_\varepsilon$ for $\Gamma'(f_\varepsilon/M)$, this equation is

$$\partial_t(g_\varepsilon\gamma_\varepsilon) + \frac{1}{\varepsilon}v \cdot \nabla_x(g_\varepsilon\gamma_\varepsilon) = \frac{1}{\varepsilon^3}\hat{\gamma}_\varepsilon\frac{1}{M}Q(f_\varepsilon, f_\varepsilon).$$

Multiplying each size of the equation above by $\varphi(v)\mathbf{1}_{|v|^2 \le K_\varepsilon}$ with $K_\varepsilon = K|\log\varepsilon|$ for some K to be fixed later and φ some collision invariant, then averaging with respect to Mdv leads to

$$\partial_t \int Mg_\varepsilon\gamma_\varepsilon\varphi\mathbf{1}_{|v|^2 \le K_\varepsilon}dv + \frac{1}{\varepsilon}\nabla_x \cdot \int Mg_\varepsilon\gamma_\varepsilon\varphi\mathbf{1}_{|v|^2 \le K_\varepsilon}vdv$$
$$= \frac{1}{\varepsilon^3}\int \hat{\gamma}_\varepsilon Q(f_\varepsilon, f_\varepsilon)\varphi\mathbf{1}_{|v|^2 \le K_\varepsilon}dv.$$

Below we use the notations

$$F_\varepsilon(\Phi) = \frac{1}{\varepsilon} \int Mg_\varepsilon\gamma_\varepsilon\Phi 1_{|v|^2 \le K_\varepsilon} dv, \quad F_\varepsilon(\Psi) = \frac{1}{\varepsilon} \int Mg_\varepsilon\gamma_\varepsilon\Psi 1_{|v|^2 \le K_\varepsilon} dv \quad (4.22)$$

for the fluxes (where Φ and Ψ are the kinetic momentum and energy fluxes defined by (3.17)), and

$$\begin{aligned} D_\varepsilon(v) &= \frac{1}{\varepsilon^3} \int \hat{\gamma}_\varepsilon Q(f_\varepsilon, f_\varepsilon) v 1_{|v|^2 \le K_\varepsilon} dv, \\ D_\varepsilon\left(\frac{1}{2}(|v|^2 - 5)\right) &= \frac{1}{\varepsilon^3} \int \hat{\gamma}_\varepsilon Q(f_\varepsilon, f_\varepsilon)\frac{1}{2}(|v|^2 - 5) 1_{|v|^2 \le K_\varepsilon} dv \end{aligned} \quad (4.23)$$

for the corresponding *conservation defects*.

The Navier-Stokes motion equation is then obtained by passing to the limit as $\varepsilon \to 0$ modulo gradient fields in

$$\partial_t \int Mg_\varepsilon\gamma_\varepsilon 1_{|v|^2 \le K_\varepsilon} v dv + \nabla_x \cdot F_\varepsilon(\Phi) + \nabla_x p_\varepsilon = D_\varepsilon(v), \quad (4.24)$$

while the temperature equation is obtained by passing to the limit in

$$\partial_t \int Mg_\varepsilon\gamma_\varepsilon 1_{|v|^2 \le K_\varepsilon}\frac{1}{2}(|v|^2 - 5) dv + \nabla_x \cdot F_\varepsilon(\Psi) = D_\varepsilon\left(\frac{1}{2}(|v|^2 - 5)\right). \quad (4.25)$$

The first step of the proof is then to establish the vanishing of conservation defects :

Proposition 4.2.1 *Under the same assumptions as in Theorem 4.1.6, one has*

$$D_\varepsilon(1) \to 0, \quad D_\varepsilon(v) \to 0 \text{ and } D_\varepsilon\left(\frac{1}{2}(|v|^2 - 5)\right) \to 0 \text{ in } L^1_{loc}(dtdx) \text{ as } \varepsilon \to 0.$$

Proof. In order to establish the previous proposition, we will introduce a convenient decomposition of $D_\varepsilon(\varphi)$ for any collision invariant φ, then estimate the different terms using the bounds (4.18), (4.19), and (4.20), as well as the following equiintegrability statement

$$(1 + |v|^p)M|\hat{g}_\varepsilon|^2 \text{ is uniformly integrable on } [0, T] \times K \times \mathbf{R}^3, \quad (4.26)$$

for all $T > 0$, $K \subset\subset \Omega$ and $p < 2$, coming from Proposition 3.3.5, and which will be proved later (see Proposition 4.3.1).

The decomposition to be used is as follows

$$\begin{aligned} D_\varepsilon(\varphi) &= \frac{1}{\varepsilon^3} \iiint \hat{\gamma}_\varepsilon 1_{|v|^2 \le K_\varepsilon}(\sqrt{f'_\varepsilon f'_{\varepsilon*}} - \sqrt{f_\varepsilon f_{\varepsilon*}})^2 \varphi B(v - v_*, \sigma) dv dv_* d\sigma \\ &+ \frac{2}{\varepsilon^3} \iiint \hat{\gamma}_\varepsilon 1_{|v|^2 \le K_\varepsilon}(\sqrt{f'_\varepsilon f'_{\varepsilon*}} - \sqrt{f_\varepsilon f_{\varepsilon*}})\sqrt{f_\varepsilon f_{\varepsilon*}}\varphi B(v - v_*, \sigma) dv dv_* d\sigma \end{aligned}$$

or equivalently

$$
\begin{aligned}
D_\varepsilon(\varphi) = {} & \frac{1}{\varepsilon^3} \iiint \hat{\gamma}_\varepsilon \mathbf{1}_{|v|^2 \le K_\varepsilon} (\sqrt{f'_\varepsilon f'_{\varepsilon *}} - \sqrt{f_\varepsilon f_{\varepsilon *}})^2 \varphi B(v - v_*, \sigma) dv dv_* d\sigma \\
& - \frac{2}{\varepsilon^3} \iiint \hat{\gamma}_\varepsilon \mathbf{1}_{|v|^2 > K_\varepsilon} (\sqrt{f'_\varepsilon f'_{\varepsilon *}} - \sqrt{f_\varepsilon f_{\varepsilon *}}) \sqrt{f_\varepsilon f_{\varepsilon *}} \varphi B(v - v_*, \sigma) dv dv_* d\sigma \\
& + \frac{2}{\varepsilon^3} \iiint \hat{\gamma}_\varepsilon (1 - \hat{\gamma}_{\varepsilon *})(\sqrt{f'_\varepsilon f'_{\varepsilon *}} - \sqrt{f_\varepsilon f_{\varepsilon *}}) \sqrt{f_\varepsilon f_{\varepsilon *}} \varphi B(v - v_*, \sigma) dv dv_* d\sigma \\
& + \frac{2}{\varepsilon^3} \iiint \hat{\gamma}_\varepsilon \hat{\gamma}_{\varepsilon *} (1 - \hat{\gamma}'_\varepsilon \hat{\gamma}'_{\varepsilon *})(\sqrt{f'_\varepsilon f'_{\varepsilon *}} - \sqrt{f_\varepsilon f_{\varepsilon *}}) \sqrt{f_\varepsilon f_{\varepsilon *}} \varphi B(v - v_*, \sigma) dv dv_* d\sigma \\
& - \frac{1}{2\varepsilon^3} \iiint \hat{\gamma}_\varepsilon \hat{\gamma}'_\varepsilon \hat{\gamma}_{\varepsilon *} \hat{\gamma}'_{\varepsilon *} (\sqrt{f'_\varepsilon f'_{\varepsilon *}} - \sqrt{f_\varepsilon f_{\varepsilon *}})^2 (\varphi + \varphi_*) B(v - v_*, \sigma) dv dv_* d\sigma \\
& \overset{\text{def}}{=} D^1_\varepsilon(\varphi) + D^2_\varepsilon(\varphi) + D^3_\varepsilon(\varphi) + D^4_\varepsilon(\varphi) + D^5_\varepsilon(\varphi)
\end{aligned}
\tag{4.27}
$$

where we have used that φ is a collision invariant to symmetrize the last term.

• That the term $D^1_\varepsilon(\varphi)$ vanishes for $\varphi(v) = O(|v|^2)$ as $|v| \to +\infty$ is easily seen as follows

$$
\begin{aligned}
\|D^1_\varepsilon(\varphi)\|_{L^1_{t,x}} & \le C\varepsilon \|\hat{\gamma}_\varepsilon \mathbf{1}_{|v|^2 \le K_\varepsilon} \varphi\|_{L^\infty} \left\| \frac{\sqrt{f'_\varepsilon f'_{\varepsilon *}} - \sqrt{f_\varepsilon f_{\varepsilon *}}}{\varepsilon^2} \right\|^2_{L^2(dt dx B dv dv_* d\sigma)} \\
& = O(\varepsilon |\log \varepsilon|)
\end{aligned}
$$

because of the entropy dissipation bound.

• The second term $D^2_\varepsilon(\varphi)$ is controlled by the following estimate on the tails of Gaussian distributions :

$$
\int_{|v|^2 > R} |v|^p M(v) dv \sim \sqrt{\frac{2}{\pi}} R^{(p+1)/2} e^{-R/2}
\tag{4.28}
$$

We have indeed, because of the upper bound on the collision cross-section in (2.8)

$$
\begin{aligned}
& |D^2_\varepsilon(\varphi)| \\
& \le \frac{2}{\varepsilon} \left\| \frac{\sqrt{f'_\varepsilon f'_{\varepsilon *}} - \sqrt{f_\varepsilon f_{\varepsilon *}}}{\varepsilon^2} \right\|_{L^2(B dv dv_* d\sigma)} \left\| \hat{\gamma}_\varepsilon \mathbf{1}_{|v|^2 > K_\varepsilon} \varphi \sqrt{f_\varepsilon f_{\varepsilon *}} \right\|_{L^2(B dv dv_* d\sigma)} \\
& \le \frac{2}{\varepsilon} \left\| \frac{\sqrt{f'_\varepsilon f'_{\varepsilon *}} - \sqrt{f_\varepsilon f_{\varepsilon *}}}{\varepsilon^2} \right\|_{L^2(B dv dv_* d\sigma)} \left\| \hat{\gamma}_\varepsilon \sqrt{\frac{f_\varepsilon}{M}} \right\|_{L^\infty} \left\| \sqrt{\frac{f_\varepsilon}{M}} \right\|_{L^2((1+|v|)^\beta M dv)} \\
& \quad \times \|\mathbf{1}_{|v|^2 > K_\varepsilon} \varphi\|_{L^2((1+|v|)^\beta M dv)}
\end{aligned}
$$

where β is the parameter arising in Grad's cut-off assumption (2.8).

Thus, using the entropy dissipation bound, some pointwise estimate on $\sqrt{z}\gamma(z)$, the Gaussian decay estimate (4.28) and the $L^\infty_t(L^1_{loc}(dx, L^1(M(1 + |v|^2)dv)))$ bound on the fluctuation coming from Young's inequality (see Lemma 3.1.2), we get for all $\varphi(v) = O(|v|^2)$ as $|v| \to +\infty$

$$\|D_\varepsilon^2(\varphi)\|_{L_t^2(L_x^1)} = O(\varepsilon^{K/2-1}|\log\varepsilon|^{(\beta+5)/2}) \to 0$$

as soon as $K > 2$.

- The last term $D_\varepsilon^5(\varphi)$ is mastered using the same tools.

For high energies, i.e. when $|v|^2 + |v_*|^2 > K|\log\varepsilon|$, we use a variant of the estimate (4.28) on the tails of Gaussian distributions, namely

$$\iint_{|v|^2+|v_*|^2>R} (|v|^2 + |v_*|^2)^{p/2} M(v)M(v_*)dvdv_* \sim CR^{(p+4)/2}e^{-R/2},$$

to obtain, using Grad's cut off assumption (2.8),

$$
\begin{aligned}
|D_\varepsilon^{5>}(\varphi)| \\
\overset{\text{def}}{=} \left| \frac{1}{2\varepsilon^3} \iiint \hat\gamma_\varepsilon\hat\gamma_\varepsilon'\hat\gamma_{\varepsilon*}\hat\gamma_{\varepsilon*}' \mathbf{1}_{|v|^2+|v_*|^2>K_\varepsilon}(\sqrt{f_\varepsilon'f_{\varepsilon*}'} - \sqrt{f_\varepsilon f_{\varepsilon*}})^2(\varphi+\varphi_*)Bdvdv_*d\sigma \right| \\
\leq \frac{2}{\varepsilon^3}\left\| \frac{f_\varepsilon}{M}\hat\gamma_\varepsilon \right\|_{L^\infty}^2 \|\hat\gamma_\varepsilon\|_{L^\infty}^2 \left\| \mathbf{1}_{|v|^2+|v_*|^2>K_\varepsilon}(\varphi+\varphi_*) \right\|_{L^1(MM_*Bdvdv_*d\sigma)} \\
\leq \frac{C}{\varepsilon^3} \iint_{|v|^2+|v_*|^2>K_\varepsilon}(|v|^2+|v_*|^2)^{1+\beta/2}M(v)M(v_*)dvdv_*
\end{aligned}
$$

so that

$$\|D_\varepsilon^{5>}(\varphi)\|_{L_{t,x}^\infty} = O(\varepsilon^{K/2-3}|\log\varepsilon|^{(\beta+6)/2}) \to 0$$

for all $\varphi(v) = O(|v|^2)$ as $|v| \to +\infty$, as soon as $K > 6$.

For moderated energies, i.e. when $|v|^2 + |v_*|^2 \leq K|\log\varepsilon|$, the entropy dissipation bound provides (as in the first step above)

$$\|D_\varepsilon^{5<}(\varphi)\|_{L_{t,x}^1} = O(\varepsilon|\log\varepsilon|)$$

for $\varphi(v) = O(|v|^2)$ as $|v| \to +\infty$.

- To handle $D_\varepsilon^3(\varphi)$ requires the additional equiintegrability statement (4.26). For each $T > 0$ and each compact $K \subset \Omega$, one has by Cauchy-Schwarz' inequality

$$
\begin{aligned}
\|D_\varepsilon^3(\varphi)\|_{L^1([0,T]\times K)} \leq C \left\| \frac{\sqrt{f_\varepsilon'f_{\varepsilon*}'} - \sqrt{f_\varepsilon f_{\varepsilon*}}}{\varepsilon^2} \right\|_{L^2(dtdxBdvdv_*d\sigma)} \|\varphi\|_{L^2((1+|v|)^\beta Mdv)} \\
\times \left\| \hat\gamma_\varepsilon\sqrt{\frac{f_\varepsilon}{M}} \right\|_{L^\infty} \left\| \frac{1}{\varepsilon}(1-\hat\gamma_\varepsilon)\sqrt{\frac{f_\varepsilon}{M}} \right\|_{L^2([0,T]\times K,L^2((1+|v|)^\beta Mdv))} \\
\leq C \left\| \frac{1}{\varepsilon}(1-\hat\gamma_\varepsilon)(1+\frac{\varepsilon}{2}\hat g_\varepsilon) \right\|_{L^2([0,T]\times K,L^2((1+|v|)^\beta Mdv))}
\end{aligned}
$$

Because of the condition on the support of $1 - \gamma$

$$\frac{1}{\varepsilon}|1 - \hat\gamma_\varepsilon|(1 + \varepsilon|\hat g_\varepsilon|) \leq 3|1 - \hat\gamma_\varepsilon||\hat g_\varepsilon|$$

In particular, from (4.26) and the fact that $1 - \hat{\gamma}_\varepsilon$ is bounded in L^∞ and converges a.e. to 0, we deduce by the Product Limit Theorem (stated in Appendix A) that for any $p < 2$

$$\frac{1}{\varepsilon}(1 - \hat{\gamma}_\varepsilon)(1 + \varepsilon|\hat{g}_\varepsilon|) \to 0 \text{ in } L^2([0,T] \times K, L^2((1 + |v|^p)M dv)). \qquad (4.29)$$

Therefore

$$\|D_\varepsilon^3(\varphi)\|_{L^1([0,T] \times K)} \to 0 \text{ as } \varepsilon \to 0.$$

- A similar argument provides the convergence of the remaining term $D_\varepsilon^4(\varphi)$. For each $T > 0$ and each compact $K \subset \Omega$, one has by Cauchy-Schwarz inequality

$$\|D_\varepsilon^4(\varphi)\|_{L^1([0,T] \times K)} \leq C \left\| \frac{\sqrt{f_\varepsilon' f_{\varepsilon*}'} - \sqrt{f_\varepsilon f_{\varepsilon*}}}{\varepsilon^2} \right\|_{L^2(dtdxBdvdv_*d\sigma)} \left\| \hat{\gamma}_\varepsilon \sqrt{\frac{f_\varepsilon}{M}} \right\|_{L^\infty}^2$$

$$\times \left\| \frac{1}{\varepsilon}(1 - \hat{\gamma}_\varepsilon \hat{\gamma}_{\varepsilon*})\varphi \right\|_{L^2([0,T] \times K, L^2(MM_*Bdvdv_*d\sigma))}$$

$$\leq C \left\| \frac{1}{\varepsilon}(1 - \hat{\gamma}_\varepsilon) \right\|_{L^2([0,T] \times K, L^p(Mdv))}$$

for any $p > 2$.

Using again the pointwise estimate

$$\frac{1}{\varepsilon}|1 - \gamma_\varepsilon| \leq 3|\hat{g}_\varepsilon||1 - \gamma_\varepsilon|$$

$$\leq 3(|\Pi \hat{g}_\varepsilon| + |\hat{g}_\varepsilon - \Pi \hat{g}_\varepsilon|)|1 - \gamma_\varepsilon|$$

with (4.19), we obtain that for all $p < +\infty$

$$\frac{1 - \gamma_\varepsilon}{\varepsilon} = O(1)_{L_t^\infty(L^2(dx, L^p(Mdv)))} + O(\varepsilon)_{L_{loc}^1(dtdx, L^2(Mdv))}$$

On the other hand,

$$\frac{1 - \gamma_\varepsilon}{\varepsilon} = O(1)_{L_t^\infty(L^2(dxMdv))} \text{ and } \frac{1 - \gamma_\varepsilon}{\varepsilon} = O\left(\frac{1}{\varepsilon}\right)_{L_{t,x,v}^\infty}$$

Thus, for all $p < 4$

$$\frac{1 - \gamma_\varepsilon}{\varepsilon} = O(1)_{L_{loc}^2(dtdx, L^p(Mdv))}.$$

Using (4.29), we then get

$$\frac{1 - \gamma_\varepsilon}{\varepsilon} \to 0 \text{ in } L_{loc}^2(dtdx, L^p(Mdv)) \qquad (4.30)$$

for all $p < 4$.

We therefore conclude that

$$\|D_\varepsilon^4(\varphi)\|_{L^1([0,T] \times K)} \to 0 \text{ as } \varepsilon \to 0.$$

Plugging all estimates in (4.27) leads to the expected convergence. $\qquad \square$

4.2.3 Decomposition of the Flux Term

We have then to characterize the asymptotic behaviour of the flux terms :

Proposition 4.2.2 *Under the same assumptions as in Theorem 4.1.6, one has*

$$F_\varepsilon(\Phi) - \frac{1}{2} \int M(\Pi \hat{g}_\varepsilon)^2 \Phi dv + 2 \int M \hat{q}_\varepsilon \tilde{\Phi} dv \to 0,$$

$$F_\varepsilon(\Psi) - \frac{1}{2} \int M(\Pi \hat{g}_\varepsilon)^2 \Psi dv + 2 \int M \hat{q}_\varepsilon \tilde{\Psi} dv \to 0,$$

in $L^1_{loc}(dtdx)$ *as* $\varepsilon \to 0$.

Proof. Let us first recall from the formal derivation that the flux term should be decomposed according to a convection term and a diffusion term, using the fact that the distribution f_ε is expected to be well approximated by the corresponding local thermodynamic equilibrium :

$$f_\varepsilon = \mathcal{M}_{f_\varepsilon} + (f_\varepsilon - \mathcal{M}_{f_\varepsilon}).$$

However, we have seen in the previous chapter that such an ingenuous method fails because the entropy dissipation is not known to control the quantity $(f_\varepsilon - \mathcal{M}_{f_\varepsilon})$ in some suitable norm.

The convenient alternative is actually to consider the linearized version of this decomposition

$$\hat{g}_\varepsilon = \Pi \hat{g}_\varepsilon + (\hat{g}_\varepsilon - \Pi \hat{g}_\varepsilon)$$

where Π is the L^2 orthogonal projection on $\mathrm{Ker}\mathcal{L}_M$, and $\hat{g}_\varepsilon$ is the renormalized fluctuation (which can be studied in the framework of the linear hilbertian theory).

• The first step consists therefore in introducing a suitable decomposition of the flux term, well adapted to the structure of the collision operator.

$$
\begin{aligned}
F_\varepsilon(\zeta) &= \frac{1}{4} \int M \hat{g}_\varepsilon^2 \gamma_\varepsilon \zeta \mathbf{1}_{|v|^2 \leq K_\varepsilon} dv + \frac{1}{\varepsilon} \int M \hat{g}_\varepsilon \gamma_\varepsilon \zeta \mathbf{1}_{|v|^2 \leq K_\varepsilon} dv \\
&= \frac{1}{4} \int M(\Pi \hat{g}_\varepsilon)^2 \zeta dv + \frac{1}{\varepsilon} \int M \hat{g}_\varepsilon \zeta dv \\
&\quad + \frac{1}{4} \int M(\hat{g}_\varepsilon^2 - (\Pi \hat{g}_\varepsilon)^2) \gamma_\varepsilon \mathbf{1}_{|v|^2 \leq K_\varepsilon} \zeta dv + \frac{1}{4} \int M(\gamma_\varepsilon \mathbf{1}_{|v|^2 \leq K_\varepsilon} - 1)(\Pi \hat{g}_\varepsilon)^2 \zeta dv \\
&\quad + \frac{1}{\varepsilon} \int M \hat{g}_\varepsilon (\gamma_\varepsilon \mathbf{1}_{|v|^2 \leq K_\varepsilon} - 1) \zeta dv \\
&\stackrel{\mathrm{def}}{=} \frac{1}{4} \int M(\Pi \hat{g}_\varepsilon)^2 \zeta dv + \frac{1}{\varepsilon} \int M \hat{g}_\varepsilon \zeta dv + F^1_\varepsilon(\zeta) + F^2_\varepsilon(\zeta) + F^3_\varepsilon(\zeta)
\end{aligned}
$$

By (4.19), (4.26) and (4.29) the remainder terms $F^1_\varepsilon(\zeta)$, $F^2_\varepsilon(\zeta)$ and $F^3_\varepsilon(\zeta)$ are expected to converge to 0. It remains then to determine the asymptotic behaviour of the second term in the right hand side of the previous identity.

The crucial remark here is that ζ belongs to $(\mathrm{Ker}\mathcal{L}_M)^\perp$ so that there exists a (unique) element $\tilde{\zeta} \in L^2(\nu M dv) \cap (\mathrm{Ker}\mathcal{L}_M)^\perp$ such that $\mathcal{L}_M\tilde{\zeta} = \zeta$. By Proposition 3.2.2 and the identity

$$M\tilde{\mathcal{L}}_M \hat{g}_\varepsilon = \frac{\varepsilon}{2}\tilde{Q}(M\hat{g}_\varepsilon, M\hat{g}_\varepsilon) - \frac{2}{\varepsilon}\tilde{Q}(\sqrt{Mf_\varepsilon}, \sqrt{Mf_\varepsilon})$$

one has therefore

$$
\begin{aligned}
\frac{1}{\varepsilon}\int M\hat{g}_\varepsilon \zeta dv &= \frac{1}{\varepsilon}\int M\mathcal{L}_M \hat{g}_\varepsilon \tilde{\zeta} dv \\
&= \frac{1}{2}\int Q(M\hat{g}_\varepsilon, M\hat{g}_\varepsilon)\tilde{\zeta} dv - 2\int M\hat{q}_\varepsilon \tilde{\zeta} dv \\
&= \frac{1}{2}\int Q(M\Pi\hat{g}_\varepsilon, M\Pi\hat{g}_\varepsilon)\tilde{\zeta} dv - 2\int M\hat{q}_\varepsilon \tilde{\zeta} dv \\
&\quad + \frac{1}{2}\int Q\big(M(\hat{g}_\varepsilon - \Pi\hat{g}_\varepsilon), M(\hat{g}_\varepsilon + \Pi\hat{g}_\varepsilon)\big)\tilde{\zeta} dv \\
&\overset{\text{def}}{=} \frac{1}{2}\int Q(M\Pi\hat{g}_\varepsilon, M\Pi\hat{g}_\varepsilon)\tilde{\zeta} dv - 2\int M\hat{q}_\varepsilon \tilde{\zeta} dv + F_\varepsilon^4(\zeta)
\end{aligned}
$$

Combining both equalities and using identity (4.16), we finally deduce that

$$F_\varepsilon(\zeta) - \frac{1}{2}\int M(\Pi\hat{g}_\varepsilon)^2 \zeta dv + 2\int M\hat{q}_\varepsilon \tilde{\zeta} dv \tag{4.31}$$
$$= F_\varepsilon^1(\zeta) + F_\varepsilon^2(\zeta) + F_\varepsilon^3(\zeta) + F_\varepsilon^4(\zeta)$$

so that proving Proposition 4.2.2 comes down to establish the convergence to zero of the four remainder terms $F_\varepsilon^1(\zeta)$, $F_\varepsilon^2(\zeta)$, $F_\varepsilon^3(\zeta)$ and $F_\varepsilon^4(\zeta)$.

• The first term $F_\varepsilon^1(\zeta)$ requires a careful treatment because of the weight $\zeta(v) = O(|v|^3)$ as $|v| \to +\infty$. By the Cauchy-Schwarz inequality, for each $T > 0$ and each compact $K \subset \Omega$

$$\|F_\varepsilon^1(\zeta)\|_{L^1([0,T]\times K)} \leq \big\|(\hat{g}_\varepsilon + \Pi\hat{g}_\varepsilon)\gamma_\varepsilon \zeta \mathbb{1}_{|v|^2 \leq K|\log \varepsilon|}\big\|_{L^2([0,T]\times K, L^2(Mdv))}$$
$$\times \|\hat{g}_\varepsilon - \Pi\hat{g}_\varepsilon\|_{L^2([0,T]\times K, L^2(Mdv))}$$

By (4.26) and the relaxation bound (4.19), one easily establishes, using Lebesgue's theorem, that for any $q < 2$

$$\|\hat{g}_\varepsilon - \Pi\hat{g}_\varepsilon\|_{L^2([0,T]\times K, L^2((1+|v|^q)Mdv))} \to 0 \quad \text{as } \varepsilon \to 0.$$

It remains then to obtain a suitable control on large velocities. By (4.18), one has for all $p < +\infty$

$$\Pi\hat{g}_\varepsilon = O(1)_{L_t^\infty(L_x^2(L^p(Mdv)))}$$

so that

$$\hat{g}_\varepsilon = O(1)_{L_t^\infty(L^2(dx, L^p(Mdv)))} + O(\varepsilon)_{L_{loc}^1(dtdx, L^2(Mdv))}$$

On the other hand,

$$\hat{g}_\varepsilon \gamma_\varepsilon^2 = O(1)_{L_t^\infty(L^2(dxMdv))} \text{ and } \hat{g}_\varepsilon \gamma_\varepsilon^2 = O\left(\frac{1}{\varepsilon}\right)_{L_{t,x,v}^\infty}$$

Thus, for all $p < 4$

$$(\hat{g}_\varepsilon + \Pi\hat{g}_\varepsilon)\gamma_\varepsilon = O(1)_{L_{loc}^2(dtdx, L^p(Mdv))}$$

Plugging this last estimate in the inequality leads to

$$F_\varepsilon^1(\zeta) \to 0 \text{ in } L_{loc}^1(dtdx) \text{ as } \varepsilon \to 0.$$

• The term $F_\varepsilon^2(\zeta)$ is easily disposed of, using the equiintegrability statement (4.26) which implies in particular that

$$M(\Pi\hat{g}_\varepsilon)^2(1 + |v|^p) \text{ is uniformly integrable on } [0, T] \times K \times \mathbf{R}^3,$$

for each $T > 0$, each compact $K \subset \Omega$ and each $p < +\infty$. Then, by the Product Limit Theorem (stated in Appendix A), as $(\gamma_\varepsilon \mathbf{1}_{|v|^2 \le K_\varepsilon} - 1)$ is bounded in L^∞ and converges a.e. to 0,

$$F_\varepsilon^2(\zeta) \to 0 \text{ in } L_{loc}^1(dtdx) \text{ as } \varepsilon \to 0.$$

• In order to get the convergence of $F_\varepsilon^3(\zeta)$, we use both the estimate (4.28) on the tails of Gaussian distributions, and the convergence (4.30) obtained in the previous paragraph.

One indeed has, by (4.28),

$$\left\| \frac{1}{\varepsilon} \int M\hat{g}_\varepsilon \gamma_\varepsilon (\mathbf{1}_{|v|^2 \le K_\varepsilon} - 1)\zeta dv \right\|_{L_t^\infty(L^2(dx))}$$
$$\le \frac{1}{\varepsilon} \|\gamma_\varepsilon\|_{L^\infty} \|\hat{g}_\varepsilon\|_{L_t^\infty(L^2(dxMdv))} \left(\int M\mathbf{1}_{|v|^2 > K_\varepsilon} \zeta^2 dv \right)^{1/2} \le C\varepsilon^{K/4-1} |\log\varepsilon|^{7/4}$$

since $\zeta^2(v) = O(|v|^6)$ as $|v| \to +\infty$.

On the other hand, by (4.30)

$$\left\| \frac{1}{\varepsilon} \int M\hat{g}_\varepsilon(\gamma_\varepsilon - 1)\zeta dv \right\|_{L^2([0,T], L^1(K))}$$
$$\le \left\| \zeta \frac{\gamma_\varepsilon - 1}{\varepsilon} \right\|_{L^2([0,T] \times K, L^2(Mdv))} \|\hat{g}_\varepsilon\|_{L_t^\infty(L^2(dxMdv))} \to 0.$$

Thus,

$$F_\varepsilon^3(\zeta) \to 0 \text{ in } L_{loc}^1(dtdx) \text{ as } \varepsilon \to 0.$$

• The continuity of Q

$$\left\|\frac{1}{M}Q(Mg, Mg)\right\|_{L^2(M\nu^{-1}dv)} \leq C\|g\|_{L^2(Mdv)}\|g\|_{L^2(M(1+|v|)^\beta dv)}$$

implies that, for each $T > 0$ and each compact $K \subset \Omega$,

$$\|F_\varepsilon^4(\zeta)\|_{L^1([0,T]\times K)} \leq C\|\tilde\zeta\|_{L^2(\nu Mdv)}\|\hat g_\varepsilon - \Pi\hat g_\varepsilon\|_{L^2([0,T]\times K, L^2((1+|v|)^\beta Mdv))}$$

$$\times \|\hat g_\varepsilon + \Pi\hat g_\varepsilon\|_{L^2([0,T]\times K, L^2((1+|v|)^\beta Mdv))}.$$

Therefore

$$F_\varepsilon^4(\zeta) \to 0 \text{ in } L^1_{loc}(dtdx) \text{ as } \varepsilon \to 0.$$

Combining all results leads to the expected convergence. □

4.3 Study of the Convection and Diffusion Terms

Let us first recall that the strategy used here to obtain the Navier-Stokes Fourier equations consists in taking limits in (4.24) (modulo gradients) and (4.25), namely in

$$\partial_t \int Mg_\varepsilon\gamma_\varepsilon \mathbf{1}_{|v|^2\leq K_\varepsilon} vdv + \nabla_x \cdot F_\varepsilon(\Phi) + \nabla_x p_\varepsilon = D_\varepsilon(v),$$

$$\partial_t \int Mg_\varepsilon\gamma_\varepsilon \mathbf{1}_{|v|^2\leq K_\varepsilon} \frac{1}{2}(|v|^2 - 5)dv + \nabla_x \cdot F_\varepsilon(\Psi) = D_\varepsilon\left(\frac{1}{2}(|v|^2 - 5)\right).$$

Therefore, in view of the results established in the previous section, our main task now is to determine the asymptotic behaviour of the convection terms

$$\frac{1}{2}\int M(\Pi\hat g_\varepsilon)^2\Phi dv \text{ and } \frac{1}{2}\int M(\Pi\hat g_\varepsilon)^2\Psi dv,$$

and of the diffusion terms

$$2\int M\hat q_\varepsilon\tilde\Phi dv \text{ and } 2\int M\hat q_\varepsilon\tilde\Psi dv.$$

Explicit computations show that the convection terms can be expressed in terms of the moments of $\Pi\hat g_\varepsilon$ (which are equal by definition to those of $\hat g_\varepsilon$). We have indeed (see [5] p. 71 for instance)

$$\frac{1}{2}\int M(\Pi\hat g_\varepsilon)^2\Phi dv = u_\varepsilon^{\otimes 2} - \frac{1}{3}|u_\varepsilon|^2 \text{ Id},$$

$$\frac{1}{2}\int M(\Pi\hat g_\varepsilon)^2\Psi dv = \frac{5}{2}u_\varepsilon\theta_\varepsilon,$$

where u_ε and θ_ε are the bulk velocity and temperature associated with $\hat{g}_\varepsilon$. The difficulty is therefore to obtain enough strong compactness on the moments of $\hat{g}_\varepsilon$ to establish the convergence of the previous quadratic quantities.

Dealing with the diffusion terms is less difficult, in the sense that it involves only weak compactness arguments. The only point is to characterize the weak limit of the sequence $(\hat{q}_\varepsilon)$.

4.3.1 Spatial Regularity Coming from Averaging Lemmas

The first step is then to study the spatial regularity of the renormalized fluctuation $\hat{g}_\varepsilon$, and of the corresponding moments, which is based on the properties of the free transport operator stated in the previous chapter, namely the dispersion and velocity averaging results.

Proposition 4.3.1 *Under the same assumptions as in Theorem 4.1.6, one has for all $p < 2$, all $T > 0$ and all compact $K \subset \Omega$*
- *the family $((1 + |v|^p)M|\hat{g}_\varepsilon|^2)$ is uniformly integrable on $[0,T] \times K \times \mathbf{R}^3$,*
- *for any $\varphi \in L^2(Mdv)$*

$$\left\| \int M\hat{g}_\varepsilon(t, x + y, v)\varphi(v)dv - \int M\hat{g}_\varepsilon(t, x, v)\varphi(v)dv \right\|^2_{L^2([0,T] \times K)} \to 0 \text{ as } |y| \to 0.$$

Proof. Starting from the uniform equiintegrability in v stated in (4.20), we expect these estimates to be obtained thanks to the mixing and velocity averaging properties of the free transport operator described in the previous chapter.

- The first requirement is therefore to get a control on the source term. As the squareroot renormalization is not admissible for the Boltzmann equation (because of the singularity at 0), we introduce some modified renormalized fluctuation

$$\frac{\sqrt{f_\varepsilon/M + \varepsilon^a} - 1}{\varepsilon},$$

for some $a > 0$ to be chosen later, and compute

$$(\varepsilon\partial_t + v \cdot \nabla_x)\frac{\sqrt{f_\varepsilon/M + \varepsilon^a} - 1}{\varepsilon}.$$

We will prove that

$$(\varepsilon\partial_t + v \cdot \nabla_x)\frac{\sqrt{f_\varepsilon/M + \varepsilon^a} - 1}{\varepsilon}$$
$$= O(\varepsilon^{2-a/2})_{L^1(dtdxMdv)} + O(1)_{L^2(dtdx\nu^{-1}Mdv)} + O(\varepsilon)_{L^1_{loc}(dtdx,L^2((1+|v|)^{-\beta}Mdv))} \tag{4.32}$$

Renormalize the Boltzmann equation (4.10) relatively to M by the non-linearity $\Gamma_\varepsilon(z) = \frac{1}{\varepsilon}(\sqrt{z + \varepsilon^a} - 1)$

$$(\varepsilon\partial_t + v \cdot \nabla_x)\frac{\sqrt{f_\varepsilon/M + \varepsilon^a} - 1}{\varepsilon}$$

$$= \frac{1}{2\varepsilon^2} \frac{1}{\sqrt{f_\varepsilon + \varepsilon^a M}\sqrt{M}} \iint \left(\sqrt{f'_\varepsilon f'_{\varepsilon*}} - \sqrt{f_\varepsilon f_{\varepsilon*}}\right)^2 B(v - v_*, \sigma)d\sigma dv_*$$

$$+ \frac{1}{\varepsilon^2} \frac{\sqrt{f_\varepsilon}}{\sqrt{f_\varepsilon + \varepsilon^a M}\sqrt{M}} \iint \left(\sqrt{f'_\varepsilon f'_{\varepsilon*}} - \sqrt{f_\varepsilon f_{\varepsilon*}}\right)\sqrt{f_{\varepsilon*}}B(v - v_*, \sigma)d\sigma dv_*$$

$$\overset{\text{def}}{=} Q^1_\varepsilon + Q^2_\varepsilon$$

$$(4.33)$$

The L^2 bound (4.18) on $\hat{q}_\varepsilon$ implies that the first term in the right hand side satisfies

$$\|Q^1_\varepsilon\|_{L^1(dtdxMdv)} \leq \frac{1}{2}C_{in}\varepsilon^{2-a/2}.$$

The weigthed L^2 bound (4.20) on $\hat{g}_\varepsilon$ implies that

$$\sqrt{\frac{f_{\varepsilon*}}{M_*}} = 1 + O(\varepsilon)_{L^2_{loc}(dtdx, L^2(M_*(1+|v_*|)^\beta dv_*))},$$

from which we deduce that the second term in the right hand side of the free transport equation satisfies

$$Q^2_\varepsilon = O(1)_{L^2(dtdx\nu^{-1}Mdv)} + O(\varepsilon)_{L^1_{loc}(dtdx, L^2((1+|v|)^{-\beta}Mdv))}.$$

Combining both estimates leads to (4.32).

• Combining this estimate on the source term with the L^2 bound on $\hat{g}_\varepsilon$, and applying Proposition 3.3.5 of the previous chapter, we are then able to establish the equiintegrability of the following quantity

$$\phi^\delta_\varepsilon = \left(\frac{\sqrt{f_\varepsilon/M + \varepsilon^a} - 1}{\varepsilon}\right)^2 \gamma\left(\varepsilon\delta\left(\frac{\sqrt{f_\varepsilon/M + \varepsilon^a} - 1}{\varepsilon}\right)\right),$$

where γ is a smooth truncation.

Using a decomposition according to the tail of f_ε/M, one first proves that

$$\frac{\sqrt{f_\varepsilon/M + \varepsilon^a} - 1}{\varepsilon} - \hat{g}_\varepsilon = O(\varepsilon^{a-1})_{L^\infty_{t,x,v}} + O(\varepsilon^{a/2})_{L^2_{loc}(dtdx, L^2(M(1+|v|^p)dv))}$$

for $p < 2$, which coupled with the L^2 bound (4.18) on $\hat{g}_\varepsilon$ leads to

$$\left(\frac{\sqrt{f_\varepsilon/M + \varepsilon^a} - 1}{\varepsilon}\right)^2 - \hat{g}^2_\varepsilon = O(\varepsilon^{a-1})_{L^2_{loc}(dtdx, L^2((1+|v|^p)Mdv))}$$

$$+ O(\varepsilon^{a/2})_{L^2_{loc}(dtdx, L^1((1+|v|^p)Mdv))}.$$

$$(4.34)$$

In particular, by (4.20)

$$\phi^\delta_\varepsilon = O(1)_{L^1_{loc}(dtdx, L^1((1+|v|^p)Mdv))},$$

$$\phi^\delta_\varepsilon \text{ is uniformly integrable in the } v \text{ variables.}$$

$$(4.35)$$

On the other hand, from (4.32) one deduces that

$$(\varepsilon\partial_t + v \cdot \nabla_x)\phi_\varepsilon^\delta = O\left(\frac{1}{\delta}\right)_{L^1_{loc}(dtdxMdv)} \tag{4.36}$$

provided that $1 < a < 2$. Indeed

$$(\varepsilon\partial_t + v\cdot\nabla_x)\phi_\varepsilon^\delta = \frac{1}{\delta}\tilde{\gamma}\left(\delta\left(\frac{\sqrt{f_\varepsilon/M + \varepsilon^a} - 1}{\varepsilon}\right)\right)(\varepsilon\partial_t + v\cdot\nabla_x)\frac{\sqrt{f_\varepsilon/M + \varepsilon^a} - 1}{\varepsilon}$$

Thus, using (4.32), (4.20) and pointwise estimates on $\tilde{\gamma}(z) = 2z\gamma(z) + z^2\gamma'(z)$, we get (4.36).

By (4.35), (4.36) and Proposition 3.3.5, we finally get that, for all $T > 0$ and all compact $K \subset \Omega$,

$$\phi_\varepsilon^\delta \text{ is uniformly integrable on } [0, T] \times K \times \mathbf{R}^3.$$

- A simple comparison allows then to obtain the first statement in Proposition 4.3.1.

Indeed, we have

$$\left|\phi_\varepsilon^\delta - \left(\frac{\sqrt{f_\varepsilon/M + \varepsilon^a} - 1}{\varepsilon}\right)^2\right| \le \frac{C}{\varepsilon^2}\frac{f_\varepsilon}{M}\mathbf{1}_{f_\varepsilon/M > 1/\delta^2}$$

so that, by the entropy inequality,

$$\phi_\varepsilon^\delta - \left(\frac{\sqrt{f_\varepsilon/M + \varepsilon^a} - 1}{\varepsilon}\right)^2 = O\left(\frac{1}{|\log\delta|}\right)_{L^\infty_t(L^1(dxMdv))}, \tag{4.37}$$

uniformly in ε.

Combining (4.34), (4.37) and the equiintegrability statement for ϕ_ε^δ leads then to

$$M|\hat{g}_\varepsilon|^2 \text{ locally uniformly integrable on } \mathbf{R}^+ \times \Omega \times \mathbf{R}^3,$$

and we conclude using the weighted L^2 bound (4.20).

- It remains then to establish the second statement, namely the spatial compactness of the moments, which is done using again some approximation of the fluctuation and applying Theorem 3.3.6 of the previous chapter.

The estimate (4.34) shows that one can replace $\hat{g}_\varepsilon$ by

$$\Gamma_\varepsilon(f_\varepsilon/M) = \frac{\sqrt{f_\varepsilon/M + \varepsilon^a} - 1}{\varepsilon}$$

with $a > 1$ in the equicontinuity statement to be proved.

By (4.32) and the local equiintegrability of $\Gamma_\varepsilon^2(f_\varepsilon/M)$ obtained in the previous step, we obtain the equicontinuity statement in $L^1_{loc}(dtdx)$. It remains then to prove that the same convergence holds in $L^2_{loc}(dtdx)$, which is achieved by using a suitable decomposition of the integral and the equiintegrability of $\Gamma_\varepsilon^2(f_\varepsilon/M)$ (see [54] for instance). □

4.3.2 Filtering of Acoustic Waves

It remains to obtain compactness with respect to the time variable. As we shall see, the solenoidal part of $u_\varepsilon = \int M\hat{g}_\varepsilon v dv$ and the linear combination $3\theta_\varepsilon - 2\rho_\varepsilon = \int M\hat{g}_\varepsilon(|v|^2 - 5)dv$ are strongly compact, but their orthogonal complement (relatively to the incompressibility and Boussinesq constraints) are not because of high frequency oscillations in t, known as *acoustic waves*.

Nevertheless Lions and Masmoudi [76] have developed a compensated compactness argument, inspired of the filtering method found independently by Schochet [96] and Grenier [60], which allows to prove that these acoustic waves do not occur in the limiting system.

Proposition 4.3.2 *Consider two families π_ε and $\nabla\psi_\varepsilon$ uniformly bounded in $L^\infty_{loc}(\mathbf{R}^+, L^2(\Omega))$, which satisfy the following equicontinuity statement : for all compact $K \subset \Omega$ and all $T > 0$, there exists some continuity modulus ω*

$$\forall \delta \text{ sufficiently small,} \quad \begin{aligned} \|\tau_\delta\pi_\varepsilon - \pi_\varepsilon\|_{L^2([0,T]\times K)} &\leq \omega(|\delta|), \\ \|\tau_\delta\nabla\psi_\varepsilon - \nabla\psi_\varepsilon\|_{L^2([0,T]\times K)} &\leq \omega(|\delta|) \end{aligned} \tag{4.38}$$

where τ_δ denotes the spatial translation of δ. Assume that

$$\begin{aligned} \partial_t\pi_\varepsilon + \frac{1}{\varepsilon}\Delta_x\psi_\varepsilon &= \frac{1}{\varepsilon}S_\varepsilon, \\ \partial_t\nabla\psi_\varepsilon + \frac{5}{3\varepsilon}\nabla_x\pi_\varepsilon &= \frac{1}{\varepsilon}S'_\varepsilon, \end{aligned} \tag{4.39}$$

where $S_\varepsilon, S'_\varepsilon \to 0$ in $L^1_{loc}(\mathbf{R}^+, W^{-s,1}(\Omega))$ for some $s > 1$.

Then, denoting by P the Leray projection onto divergence free vector fields,

$$P\nabla_x \cdot ((\nabla\psi_\varepsilon)^{\otimes 2}) \to 0, \quad \text{and} \quad \nabla_x \cdot (\pi_\varepsilon\nabla\psi_\varepsilon) \to 0$$

in the sense of distributions on $\mathbf{R}^+ \times \Omega$.

Proof. First we introduce the following regularization : let $\chi \in C_c^\infty(\mathbf{R}^3, \mathbf{R}^+)$ be such that

$$\chi(x) = 0 \text{ if } |x| \geq 1, \text{ and } \int \chi(x)dx = 1,$$

and define

$$\chi_\delta(x) = \delta^{-3}\chi\left(\frac{x}{\delta}\right).$$

Denoting by $\pi^\delta_\varepsilon = \chi_\delta * \pi_\varepsilon$ and $\nabla\psi^\delta_\varepsilon = \chi_\delta * \pi_\varepsilon$, one has by (4.39)

$$\begin{aligned} \varepsilon\partial_t\pi^\delta_\varepsilon + \Delta_x\psi^\delta_\varepsilon &= S^\delta_\varepsilon, \\ \varepsilon\partial_t\nabla\psi^\delta_\varepsilon + \frac{5}{3}\nabla_x\pi^\delta_\varepsilon &= S'^\delta_\varepsilon, \end{aligned}$$

with $S^\delta_\varepsilon, S'^\delta_\varepsilon \to 0$ in $L^1_{loc}(\mathbf{R}^+, H^s_{loc}(\Omega))$ for all $s > 0$.

In particular, from the elementary computations

$$\nabla_x \cdot ((\nabla \psi_\varepsilon^\delta)^{\otimes 2})$$
$$= \frac{1}{2} \nabla_x (|\nabla_x \psi_\varepsilon^\delta|^2) + \nabla_x \psi_\varepsilon^\delta \Delta_x \psi_\varepsilon^\delta$$
$$= \frac{1}{2} \nabla_x (|\nabla_x \psi_\varepsilon^\delta|^2) + \nabla_x \psi_\varepsilon^\delta (S_\varepsilon^\delta - \varepsilon \partial_t \pi_\varepsilon^\delta)$$
$$= \frac{1}{2} \nabla_x (|\nabla_x \psi_\varepsilon^\delta|^2) + \nabla_x \psi_\varepsilon^\delta S_\varepsilon^\delta - \varepsilon \partial_t (\pi_\varepsilon^\delta \nabla_x \psi_\varepsilon^\delta) + \pi_\varepsilon^\delta \left(S_\varepsilon'^{\,\delta} - \frac{5}{3} \nabla_x \pi_\varepsilon^\delta \right)$$
$$= \frac{1}{2} \nabla_x \left(|\nabla_x \psi_\varepsilon^\delta|^2 - \frac{5}{3} (\pi_\varepsilon^\delta)^2 \right) - \varepsilon \partial_t (\pi_\varepsilon^\delta \nabla_x \psi_\varepsilon^\delta) + \nabla_x \psi_\varepsilon^\delta S_\varepsilon^\delta + \pi_\varepsilon^\delta S_\varepsilon'^{\,\delta}$$

and

$$\nabla_x \cdot (\pi_\varepsilon^\delta \nabla_x \psi_\varepsilon^\delta) = \pi_\varepsilon^\delta \Delta_x \psi_\varepsilon^\delta + \nabla_x \pi_\varepsilon^\delta \cdot \nabla_x \psi_\varepsilon^\delta$$
$$= \pi_\varepsilon^\delta (S_\varepsilon^\delta - \varepsilon \partial_t \pi_\varepsilon^\delta) + \nabla_x \psi_\varepsilon^\delta \cdot \frac{3}{5} (S_\varepsilon'^{\,\delta} - \varepsilon \partial_t \nabla_x \psi_\varepsilon^\delta)$$
$$= -\frac{\varepsilon}{2} \partial_t \left((\pi_\varepsilon^\delta)^2 + \frac{3}{5} |\nabla \psi_\varepsilon^\delta|^2 \right) + \pi_\varepsilon^\delta S_\varepsilon^\delta + \frac{3}{5} \nabla_x \psi_\varepsilon^\delta \cdot S_\varepsilon'^{\,\delta}$$

we deduce that, for all fixed $\delta > 0$

$$P \nabla_x \cdot ((\nabla \psi_\varepsilon^\delta)^{\otimes 2}) \to 0, \quad \text{and} \quad \nabla_x \cdot (\pi_\varepsilon^\delta \nabla \psi_\varepsilon^\delta) \to 0$$

in the sense of distributions, as $\varepsilon \to 0$.

It remains then to get rid of the regularization, which is done using the equicontinuity statement (4.38). Note that, once again, we actually use the characterization of relatively compact sets (3.28):

$$\mathcal{K} \text{ relatively compact} \quad \Leftrightarrow \quad \forall \delta > 0, \quad \exists \mathcal{K}_\delta \text{ compact}, \ d(\mathcal{K}, \mathcal{K}_\delta) \le \delta.$$

Indeed, for all compact $K \subset \Omega$ and all $T > 0$,

$$\|\nabla_x \psi_\varepsilon^\delta - \nabla_x \psi_\varepsilon\|_{L^2([0,T] \times K)} \le C \sup_{[0,\delta]} \omega, \quad \|\pi_\varepsilon^\delta - \pi_\varepsilon\|_{L^2([0,T] \times K)} \le C \sup_{[0,\delta]} \omega.$$

Therefore

$$P \nabla_x \cdot ((\nabla \psi_\varepsilon^\delta)^{\otimes 2} - (\nabla_x \psi_\varepsilon)^{\otimes 2}) \to 0, \quad \text{and} \quad \nabla_x \cdot (\pi_\varepsilon^\delta \nabla \psi_\varepsilon^\delta - \pi_\varepsilon \nabla \psi_\varepsilon) \to 0$$

in $L^1_{loc}(dtdx)$ uniformly in ε as $\delta \to 0$.

Combining both convergence statements concludes the proof. □

4.3.3 Convergence of the Nonlinear Convection Term

Equipped with the previous results, we are now able to characterize the limiting convection terms.

Proposition 4.3.3 *Under the same assumptions as in Theorem 4.1.6, one can extract a subsequence of renormalized fluctuation* $(\hat{g}_\varepsilon)$ *such that*

$$\hat{g}_\varepsilon \rightharpoonup g \quad \text{weakly in } L^2_{loc}(dt, L^2(dxMdv)).$$

Denote then by u_ε *and* θ_ε *the bulk velocity and temperature associated with the renormalized fluctuation* $\hat{g}_\varepsilon$ *and by* u *and* θ *the bulk velocity and temperature associated with* g. *Then,*

$$P\nabla_x \cdot (u_\varepsilon^{\otimes 2} - \frac{1}{3}|u_\varepsilon|^2 \,\mathrm{Id}) \to P\nabla_x \cdot (u^{\otimes 2}),$$

$$\nabla_x \cdot (u_\varepsilon \theta_\varepsilon) \to \nabla_x \cdot (u\theta),$$

in the sense of distributions.

Proof. As suggested in the previous paragraph, the convergence of the convection terms is obtained using a suitable decomposition of the moments u_ε and θ_ε in a strongly compact part, and a high frequency oscillating part. However, because the square-root renormalization of the Boltzmann equation is not admissible, we are not able to control directly the time derivatives of the moments of $\hat{g}_\varepsilon$.

• We start by approximating u_ε, θ_ε and $\rho_\varepsilon = \int M\hat{g}_\varepsilon dv$ by the moments of $g_\varepsilon \gamma_\varepsilon \mathbf{1}_{|v|^2 \leq K_\varepsilon}$:

$$\tilde{\rho}_\varepsilon = \int Mg_\varepsilon \gamma_\varepsilon \mathbf{1}_{|v|^2 \leq K_\varepsilon} dv, \quad \tilde{u}_\varepsilon = \int Mg_\varepsilon \gamma_\varepsilon \mathbf{1}_{|v|^2 \leq K_\varepsilon} v dv$$

$$\text{and } \tilde{\theta}_\varepsilon = \frac{1}{3}\int Mg_\varepsilon \gamma_\varepsilon \mathbf{1}_{|v|^2 \leq K_\varepsilon}(|v|^2 - 3)dv.$$

Observe that

$$g_\varepsilon \gamma_\varepsilon \mathbf{1}_{|v|^2 \leq K_\varepsilon} - \hat{g}_\varepsilon = \frac{1}{2}\hat{g}_\varepsilon \left(\gamma_\varepsilon \mathbf{1}_{|v|^2 \leq K_\varepsilon} \left(\sqrt{\frac{f_\varepsilon}{M}} + 1 \right) - 2 \right)$$

Using the equiintegrability statement (4.26) established in Proposition 4.3.1 and the fact that the second factor in the identity above is bounded in L^∞ and converges a.e. to 0, we obtain by the Product Limit theorem that

$$g_\varepsilon \gamma_\varepsilon \mathbf{1}_{|v|^2 \leq K_\varepsilon} - \hat{g}_\varepsilon \to 0 \text{ in } L^2_{loc}(dtdx, L^2(Mdv)). \tag{4.40}$$

In particular

$$\rho_\varepsilon - \tilde{\rho}_\varepsilon \to 0, \quad u_\varepsilon - \tilde{u}_\varepsilon \to 0, \text{ and } \theta_\varepsilon - \tilde{\theta}_\varepsilon \to 0 \text{ in } L^2_{loc}(dtdx) \text{ as } \varepsilon \to 0. \tag{4.41}$$

Furthermore, by Proposition 4.3.1, one has for all $T > 0$, all compact $K \subset \Omega$ and all $\varphi \in L^2(Mdv)$

$$\left\| \int M \hat{g}_\varepsilon(t, x+y, v)\varphi(v)dv - \int M \hat{g}_\varepsilon(t, x, v)\varphi(v)dv \right\|^2_{L^2([0,T]\times K)} \to 0$$

as $|y| \to 0$ uniformly in ε. From (4.41), we then deduce that there exists some continuity modulus ω such that, for all δ and for all ε sufficiently small

$$\|\tau_\delta \tilde{u}_\varepsilon - \tilde{u}_\varepsilon\|_{L^2([0,T]\times K)} \le \omega(\delta), \quad \|\tau_\delta \tilde{\theta}_\varepsilon - \tilde{\theta}_\varepsilon\|_{L^2([0,T]\times K)} \le \omega(\delta) \qquad (4.42)$$
$$\text{and } \|\tau_\delta \tilde{\rho}_\varepsilon - \tilde{\rho}_\varepsilon\|_{L^2([0,T]\times K)} \le \omega(\delta).$$

Finally, starting from (4.24), (4.25) and the analogous renormalized conservation of mass

$$\partial_t \int M g_\varepsilon \gamma_\varepsilon \mathbf{1}_{|v|^2 \le K_\varepsilon} dv + \frac{1}{\varepsilon} \nabla_x \cdot \int M g_\varepsilon \gamma_\varepsilon \mathbf{1}_{|v|^2 \le K_\varepsilon} v dv = D_\varepsilon(1),$$

$$\partial_t \int M g_\varepsilon \gamma_\varepsilon \mathbf{1}_{|v|^2 \le K_\varepsilon} v dv + \nabla_x \cdot F_\varepsilon(\Phi) + \frac{1}{3\varepsilon} \nabla_x \int M g_\varepsilon \gamma_\varepsilon \mathbf{1}_{|v|^2 \le K_\varepsilon} |v|^2 dv = D_\varepsilon(v),$$

$$\partial_t \int M g_\varepsilon \gamma_\varepsilon \mathbf{1}_{|v|^2 \le K_\varepsilon} \frac{1}{2}(|v|^2 - 5) dv + \nabla_x \cdot F_\varepsilon(\Psi) = D_\varepsilon(\frac{1}{2}(|v|^2 - 5)).$$

then using Propositions 4.2.1 and 4.2.2 and the L^2 bound (4.18) on $\hat{q}_\varepsilon$, we obtain

$$\partial_t \tilde{\rho}_\varepsilon + \frac{1}{\varepsilon} \nabla_x \cdot \tilde{u}_\varepsilon = D_\varepsilon(1),$$

$$\partial_t \tilde{u}_\varepsilon + \frac{1}{\varepsilon} \nabla_x (\tilde{\rho}_\varepsilon + \tilde{\theta}_\varepsilon) = D_\varepsilon(v) - \nabla_x \cdot F_\varepsilon(\Phi), \qquad (4.43)$$

$$\partial_t \frac{1}{2}(3\tilde{\theta}_\varepsilon - 2\tilde{\rho}_\varepsilon) = D_\varepsilon \left(\frac{1}{2}(|v|^2 - 5) \right) - \nabla_x \cdot F_\varepsilon(\Psi),$$

where the source terms are bounded in $L^1_{loc}(dt, W^{-1,1}_{loc}(dx))$.

• In the second step, we deal with the compact components of the moments $\tilde{u}_\varepsilon$ and $\tilde{\theta}_\varepsilon$, namely

$$P\tilde{u}_\varepsilon \text{ and } \frac{1}{2}(3\tilde{\theta}_\varepsilon - 2\tilde{\rho}_\varepsilon).$$

The equicontinuity statement (4.42) together with the fact that P is a pseudo-differential operator of order 0 implies that

$$\|\tau_\delta P\tilde{u}_\varepsilon - P\tilde{u}_\varepsilon\|_{L^2([0,T]\times K)} \to 0$$
$$\left\| \tau_\delta \left(\frac{1}{2}(3\tilde{\theta}_\varepsilon - 2\tilde{\rho}_\varepsilon) \right) - \frac{1}{2}(3\tilde{\theta}_\varepsilon - 2\tilde{\rho}_\varepsilon) \right\|_{L^2([0,T]\times K)} \to 0 \qquad (4.44)$$

uniformly in ε as $\delta \to 0$.

On the other hand, the conservation laws (4.39) imply that

$$\partial_t \frac{1}{2}(3\tilde{\theta}_\varepsilon - 2\tilde{\rho}_\varepsilon) = O(1)_{L^1_{loc}(dt, W^{-1,1}(dx))} \qquad (4.45)$$

and

$$\partial_t \int_\Omega P\tilde{u}_\varepsilon \cdot \xi dx = O(1)_{L^1_{loc}(dt)} \qquad (4.46)$$

for each compactly supported, solenoidal vector field $\xi \in H^3(\Omega)$. Also,

$$\tilde{u}_\varepsilon = O(1) \text{ in } L^\infty(\mathbf{R}_+; L^2(dx))$$

From (4.44) and (4.45) we deduce by interpolation (using for instance Aubin's lemma [2]) that

$$\frac{1}{2}(3\tilde{\theta}_\varepsilon - 2\tilde{\rho}_\varepsilon) \text{ is strongly compact in } L^2_{loc}(dtdx),$$

and we can identify its limit using the Boussinesq relation

$$\frac{1}{5}(3\tilde{\theta}_\varepsilon - 2\tilde{\rho}_\varepsilon) \to \theta \text{ strongly in } L^2_{loc}(dtdx).$$

For the solenoidal part of the velocity, the argument is more intricated because of the Leray projection. Since the class of C^∞, compactly supported solenoidal vector fields is dense in that of all L^2 solenoidal vector fields (see Appendix A of [74]), these estimates imply that

$$P\tilde{u}_\varepsilon \text{ is relatively compact in } C(\mathbf{R}_+; w\text{-}L^2(\Omega)).$$

As for the $L^2_{loc}(dtdx)$ compactness, observe that (4.44) implies that

$$P\tilde{u}_\varepsilon \star \chi_\delta \text{ is relatively compact in } L^2_{loc}(dtdx)$$

where (χ_δ) designates any mollifying sequence. Hence

$$P\tilde{u}_\varepsilon \cdot P\tilde{u}_\varepsilon \star \chi_\delta \rightharpoonup Pu \cdot Pu \star \chi_\delta$$

weakly in $L^1_{loc}(dtdx)$ as $\varepsilon \to 0$. By (4.44),

$$P\tilde{u}_\varepsilon \star \chi_\delta \to P\tilde{u}_\varepsilon$$

in $L^2_{loc}(dtdx)$ uniformly in ε as $\delta \to 0$. With this, we conclude that

$$|P\tilde{u}_\varepsilon|^2 \to |Pu|^2 \text{ in } w\text{-}L^1_{loc}(dtdx)$$

which implies in view of the incompressibility constraint that

$$P\tilde{u}_\varepsilon \to u \text{ strongly in } L^2_{loc}(dtdx).$$

In particular, one has

$$PV_x \cdot ((P\tilde{u}_\varepsilon)^{\otimes 2}) \to PV_x(u^{\otimes 2}) \text{ and } \nabla_x \cdot \left(\frac{1}{5}(3\tilde{\theta}_\varepsilon - 2\tilde{\rho}_\varepsilon)P\tilde{u}_\varepsilon\right) \to \nabla_x \cdot (\theta u)$$

$$(4.47)$$

in the sense of distributions.

• It remains then to prove that acoustic waves do not occur in the limiting equations, which is based on Proposition 4.3.2. Define indeed

$$\nabla_x \psi_\varepsilon = \tilde{u}_\varepsilon - P\tilde{u}_\varepsilon \text{ and } \pi_\varepsilon = \frac{3}{5}(\tilde{\rho}_\varepsilon + \tilde{\theta}_\varepsilon).$$

From the L^2 estimates (4.18) and (4.41), and the incompressibility and Boussinesq relations, we deduce that

$$\nabla_x \psi_\varepsilon \rightharpoonup 0, \quad \pi_\varepsilon \rightharpoonup 0 \text{ weakly in } L^2_{loc}(dtdx).$$

In particular

$$P\nabla_x \cdot (\tilde{u}_\varepsilon \otimes \nabla_x \psi_\varepsilon + \nabla_x \psi_\varepsilon \otimes \tilde{u}_\varepsilon) \to 0,$$
$$\nabla_x \cdot \left(\frac{1}{5}(3\tilde{\theta}_\varepsilon - 2\tilde{\rho}_\varepsilon)\nabla_x \psi_\varepsilon + \frac{2}{3}\pi_\varepsilon \tilde{u}_\varepsilon \right) \to 0 \tag{4.48}$$

in the sense of distributions.

By (4.42) and (4.44), $\nabla \psi_\varepsilon$ and π_ε satisfy assumption (4.38) in Proposition 4.3.2. On the other hand, from (4.43), we deduce that $\nabla \psi_\varepsilon$ and π_ε "almost satisfy" assumption (4.39) : there is actually a difficulty due to the fact that the Leray projection is non local. For a detailed treatment of that point (which requires additional regularizations), we refer to the original paper by Golse and the author [55]. Then,

$$P\nabla_x \cdot ((\nabla \psi_\varepsilon)^{\otimes 2}) \to 0 \text{ and } \nabla_x \cdot \left(\frac{2}{3}\pi_\varepsilon \nabla \psi_\varepsilon \right) \to 0 \tag{4.49}$$

in the sense of distributions.

Combining (4.47), (4.48) and (4.49) leads to

$$P\nabla_x \cdot ((\tilde{u}_\varepsilon)^{\otimes 2}) \to P\nabla_x(u^{\otimes 2}) \text{ and } \nabla_x \cdot \left(\tilde{\theta}_\varepsilon \tilde{u}_\varepsilon \right) \to \nabla_x \cdot (\theta u)$$

which, coupled with (4.41), gives the expected convergences for the convection terms. □

4.3.4 Convergence of the Diffusion Term

This paragraph is devoted to the last step in the proof of convergence inside the domain Ω, namely to the characterization of the limiting diffusion terms.

Proposition 4.3.4 *Under the same assumptions as in Theorem 4.1.6, if we denote by g and q any joint limit points of the sequences $(\hat{g}_\varepsilon)$ and $(\hat{q}_\varepsilon)$ defined by (4.17), one has*

$$q = \frac{1}{2}v \cdot \nabla_x g = \frac{1}{2}(\Phi : \nabla_x u + \Psi \cdot \nabla_x \theta) \tag{4.50}$$

where u and θ are the bulk velocity and temperature associated with the limiting fluctuation g, Φ and Ψ are the kinetic fluxes defined by (3.17), and $\tilde{\Phi}$ and $\tilde{\Psi}$ are their pseudo-inverses under $\mathcal{L}_M$.

In particular the following weak L^2 convergences hold for the diffusion terms

$$2 \int M \hat{q}_\varepsilon \tilde{\Phi} dv \to \nabla_x u : \int M \tilde{\Phi} \otimes \Phi dv = \mu(\nabla_x u + (\nabla_x u)^T),$$

$$2 \int M \hat{q}_\varepsilon \tilde{\Psi} dv \to \nabla_x \theta : \int M \tilde{\Psi} \otimes \Psi dv = \kappa \nabla_x \theta,$$

Proof. The second statement is easily deduced from the first one, using the weak compactness of $(\hat{q}_\varepsilon)$ coming from (4.18), and the structure of the kinetic flux functions Φ and Ψ and their pseudo-inverses $\tilde{\Phi}$ and $\tilde{\Psi}$ (see Remark 3.2.3). Let us then focus on the proof of the first statement.

Start from the square-root renormalization of the scaled Boltzmann equation (4.33)

$$(\varepsilon \partial_t + v \cdot \nabla_x) \frac{\sqrt{f_\varepsilon/M + \varepsilon^a} - 1}{\varepsilon}$$

$$= \frac{1}{2\varepsilon^2} \frac{1}{\sqrt{f_\varepsilon + \varepsilon^a M} \sqrt{M}} \iint \left(\sqrt{f'_\varepsilon f'_{\varepsilon *}} - \sqrt{f_\varepsilon f_{\varepsilon *}} \right)^2 B(v - v_*, \sigma) d\sigma dv_*$$

$$+ \frac{1}{\varepsilon^2} \frac{\sqrt{f_\varepsilon}}{\sqrt{f_\varepsilon + \varepsilon^a M} \sqrt{M}} \iint \left(\sqrt{f'_\varepsilon f'_{\varepsilon *}} - \sqrt{f_\varepsilon f_{\varepsilon *}} \right) \sqrt{f_{\varepsilon *}} B(v - v_*, \sigma) d\sigma dv_*$$

$$\overset{\text{def}}{=} Q^1_\varepsilon + Q^2_\varepsilon$$

By the weighted L^2 estimate (4.20), one has for all $p < 2$

$$\sqrt{\frac{f_\varepsilon}{M}} \to 1 \text{ in } L^2_{loc}(dtdx, L^2((1 + |v|^p)Mdv)) \text{ as } \varepsilon \to 0.$$

Furthermore the entropy dissipation bound shows that

$$\frac{1}{\varepsilon^2} \left(\sqrt{f'_\varepsilon f'_{\varepsilon *}} - \sqrt{f_\varepsilon f_{\varepsilon *}} \right) \text{ is weakly compact in } L^2_{loc}(dtdx, L^2(Bdvdv_*d\sigma)),$$

so that, up to extraction of a subsequence such that $\hat{q}_\varepsilon \rightharpoonup q$,

$$\frac{1}{\varepsilon^2} \frac{1}{\sqrt{M}} \iint \left(\sqrt{f'_\varepsilon f'_{\varepsilon *}} - \sqrt{f_\varepsilon f_{\varepsilon *}} \right) \sqrt{f_{\varepsilon *}} B(v - v_*, \sigma) d\sigma dv_* \rightharpoonup q$$

weakly in $L^1_{loc}(dtdx, L^1(Mdv))$. Since on the other hand $\sqrt{f_\varepsilon}/\sqrt{f_\varepsilon + \varepsilon^a M}$ is bounded in L^∞ and converges a.e. to 0, by the Product Limit theorem, we conclude that

$$Q^2_\varepsilon \rightharpoonup q \text{ weakly in } L^1_{loc}(dtdx, L^1(Mdv)).$$

On the other hand,

$$\|Q^1_\varepsilon\|_{L^1(dtdxMdv)} \leq \frac{1}{2} C_{in} \varepsilon^{2-a/2}.$$

Then using the comparison estimate (4.34) and taking limits in (4.33) leads to

$$\frac{1}{2} v \cdot \nabla_x g = q,$$

which concludes the proof. $\square$

4.4 Taking into Account Boundary Conditions

Taking limits in the boundary conditions requires that some additional difficulties be overcome :

• for the renormalized solutions to the Boltzmann equation, the trace is indeed defined in a very weak sense;

• the only uniform bound, i.e. the bound on the Darrozès-Guiraud information, gives a control rather on the quantity $(g_{\varepsilon|\partial\Omega} - \langle g_{\varepsilon}\rangle_{\partial\Omega})$ where

$$\langle g_{\varepsilon}\rangle_{\partial\Omega} = \sqrt{2\pi}\int g_{\varepsilon|\partial\Omega}M(v \cdot n(x))_+dv$$

than on the total trace $g_{\varepsilon|\partial\Omega}$.

In order to get rid of the first difficulty, a natural idea is to take limits in the weak formulation (2.31) and to prove that the moments u and θ satisfy asymptotically the weak formulation of the Navier-Stokes Fourier equations, which will be done in the case of the Navier boundary condition. In the case of the Dirichlet boundary condition, it is actually possible to take limits directly in Maxwell's boundary condition.

In both cases, the strategy initiated by Masmoudi and the author in [82] to derive the limiting boundary conditions is very similar to the one used to establish the asymptotic inside the domain. First, we check that, up to extraction of a subsequence, the traces converge in a sense to be made precise. Then, using the classical theory of traces for the transport equation recalled in the previous chapter, we show that the limiting trace is of the form

$$g_{|\partial\Omega} = u_{|\partial\Omega} \cdot v + \theta_{|\partial\Omega}\left(\frac{|v|^2 - 5}{2}\right).$$

The last and most difficult step is to take limits in Maxwell's boundary condition, either in a weak or in a strong form depending on the scaling.

In this method, a priori estimates and convergence results on the traces come both from results established inside the domain and from the boundary term in the entropy inequality. This means that, given the asymptotic behaviour of the fluctuation g_{ε} inside the domain Ω, the possibility of deriving the limiting boundary conditions depends only on the scaling of the free-transport operator. We point out in particular that such a method cannot be applied in the Euler scaling since there is no bound on the spatial derivatives of the fluctuation, which is consistent with the longstanding problem of Prandtl layers.

In all the sequel, we consider a subsequence of renormalized fluctuations $\hat{g}_{\varepsilon}$ (still denoted $\hat{g}_{\varepsilon}$) such that

$$\hat{g}_{\varepsilon} \to g \quad \text{weakly in } L^2_{loc}(dt, L^2(dxMdv))$$
$$\hat{q}_{\varepsilon} \to q \quad \text{weakly in } L^2(dtdxMdv)$$

4.4.1 A Priori Estimates Coming from the Inside

The first type of control obtained on the trace $g_{\varepsilon|\partial\Omega}$ is inferred from the bounds on g_ε and $(\varepsilon\partial_t + v \cdot \nabla_x)g_\varepsilon$ inside the domain $\mathbf{R}^+ \times \Omega \times \mathbf{R}^3$.

Proposition 4.4.1 *Under the same assumptions as in Theorem 4.1.6, the trace of the limiting fluctuation*

$$g_{|\partial\Omega} \in L^1_{loc}(dtd\sigma_x; L^1(M|v \cdot n(x)|dv))$$

satisfies the identity

$$g_{|\partial\Omega} = u_{|\partial\Omega} \cdot v + \theta_{|\partial\Omega}\left(\frac{|v|^2 - 5}{2}\right) \tag{4.51}$$

Furthermore, for all $p > 0$, and all smooth truncation $\gamma \in C^\infty(\mathbf{R}^+)$ such that

$$\gamma_{|[0,3/2]} \equiv 1, \quad |\gamma(z)| + z|\gamma'(z)| \le \frac{C}{1+z}$$

we have the following convergences :

$$(g_\varepsilon\gamma_\varepsilon)_{|\partial\Omega} \to g_{|\partial\Omega} \text{ weakly in } L^1_{loc}(dtd\sigma_x; L^1(M(1+|v|^p)|v \cdot n(x)|dv)) \tag{4.52}$$

$$\varepsilon g_{\varepsilon|\partial\Omega} \to 0 \text{ a.e. on } \mathbf{R}^+ \times \partial\Omega \times \mathbf{R}^3 \tag{4.53}$$

Proof. Proposition 4.4.1 is based on classical properties of the traces of solutions to the free transport equation (which can be found for instance in [38] and are recalled in Appendix B).

- Inside the domain Ω, we have

$$g = v \cdot u + \frac{1}{2}(|v|^2 - 5)\theta \text{ in } L^2_{loc}(dt, L^2(Mdvdx)),$$
$$v \cdot \nabla_x g = 2q = (\Phi : \nabla_x u + \Psi \cdot \nabla_x \theta) \text{ in } L^2_{loc}(dt, L^2(Mdvdx)).$$

Then, g, u and θ have traces in the classical sense, and identity (4.51) holds in $L^1_{loc}(dtd\sigma_x, L^1(M|v \cdot n(x)|dv))$.

- We have then to establish the weak compactness of the sequence $(g_\varepsilon\gamma_\varepsilon)_{|\partial\Omega}$. We recall that for any $q < 2$

$$g_\varepsilon\gamma_\varepsilon = O(1)_{L^2_{loc}(dtdx, L^2((1+|v|^q)Mdv))} \text{ and } g_\varepsilon\gamma_\varepsilon = O\left(\frac{1}{\varepsilon}\right)_{L^\infty_{t,x,v}}$$

coming from the L^2 estimate (4.18) on $\hat{g}_\varepsilon$ and some pointwise estimate on $\sqrt{z}\gamma(z)$. Furthermore, we have

$$(\varepsilon\partial_t + v \cdot \nabla_x)M(g_\varepsilon\gamma_\varepsilon)^2$$
$$= \frac{2}{\varepsilon^2} g_\varepsilon\gamma_\varepsilon\hat{\gamma}_\varepsilon Q(f_\varepsilon, f_\varepsilon)$$
$$= \frac{2}{\varepsilon^2} \iint g_\varepsilon\gamma_\varepsilon\hat{\gamma}_\varepsilon (\sqrt{f'_\varepsilon f'_{\varepsilon*}} - \sqrt{f_\varepsilon f_{\varepsilon*}})^2 B(v - v_*, \sigma)dv_*d\sigma$$
$$+ \frac{4}{\varepsilon^2} \iint g_\varepsilon\gamma_\varepsilon\hat{\gamma}_\varepsilon \sqrt{f_\varepsilon}(\sqrt{f'_\varepsilon f'_{\varepsilon*}} - \sqrt{f_\varepsilon f_{\varepsilon*}})B(v - v_*, \sigma)dv_*d\sigma$$
$$+ \frac{4}{\varepsilon^2} \iint \varepsilon g_\varepsilon\gamma_\varepsilon\hat{\gamma}_\varepsilon \sqrt{f_\varepsilon}\frac{1}{\varepsilon}(\sqrt{f_{\varepsilon*}} - 1)(\sqrt{f'_\varepsilon f'_{\varepsilon*}} - \sqrt{f_\varepsilon f_{\varepsilon*}})B(v - v_*, \sigma)dv_*d\sigma$$

where we denote as previously

$$\hat{\gamma}_\varepsilon = \hat{\gamma}\left(\frac{f_\varepsilon}{M}\right) \text{ with } \hat{\gamma}(z) = \gamma(z) + (z-1)\gamma'(z).$$

Thus using the entropy dissipation bound, we get

$$(\varepsilon\partial_t + v \cdot \nabla_x)M(g_\varepsilon\gamma_\varepsilon)^2 = O(\varepsilon)_{L^1(dtdxdv)} + O(1)_{L^1_{loc}(dtdx,L^1(dv))}.$$

The smoothness assumption made on the boundary implies the existence of a vector field $n(x)$ which belongs to $W^{1,\infty}_{loc}(\bar{\Omega})$ and coincides with the outward unit normal vector at the boundary. Then,

$$\int_{t_1}^{t_2} \int_{\partial\Omega} \int M|(g_\varepsilon\gamma_\varepsilon)_{|\partial\Omega}|^2(t, x, v)\frac{(v.n(x))^2\chi(x)}{1 + |v|^2}dv\, d\sigma_x\, dt$$
$$= \varepsilon \int_\Omega \int M\varphi|g_\varepsilon\gamma_\varepsilon|^2(t_1, x, v)dvdx - \varepsilon \int_\Omega \int M\varphi|g_\varepsilon\gamma_\varepsilon|^2(t_2, x, v)dvdx$$
$$+ \int_{t_1}^{t_2} \int_\Omega \int M(v \cdot \nabla_x\varphi)|g_\varepsilon\gamma_\varepsilon|^2(t, x, v)dvdxdt$$
$$+ \int_{t_1}^{t_2} \int_\Omega \int M\varphi(\varepsilon\partial_t + v \cdot \nabla_x)(g_\varepsilon\gamma_\varepsilon)^2(t, x, v)dvdxdt$$

with

$$\varphi(x, v) = \frac{v.n(x)\chi(x)}{(1 + |v|^2)}$$

for any localization function $\chi \in C_c^\infty(\bar{\Omega}, [0, 1])$.

From the uniform bounds on $(g_\varepsilon\gamma_\varepsilon)^2$ and $(\varepsilon\partial_t + v \cdot \nabla_x)(g_\varepsilon\gamma_\varepsilon)^2$, we deduce that

$$\int_{t_1}^{t_2} \int_{\partial\Omega} \int M|(g_\varepsilon\gamma_\varepsilon)_{|\partial\Omega}|^2(t, x, v)\frac{(v.n(x))^2\chi(x)}{1 + |v|^2}dv\, d\sigma_x\, dt \leq C \qquad (4.54)$$

The weak compactness asserted in (4.52) follows directly by Hölder's inequality. Indeed, for all $\mathcal{A} \subset [t_1, t_2] \times (\partial\Omega \cap K) \times \mathbf{R}^3$, and all $p < +\infty$,

$$\iiint_\mathcal{A} M|(\gamma_\varepsilon g_\varepsilon)_{|\partial\Omega}|(1 + |v|^p)v.n(x)|dvd\sigma_x dt$$
$$\leq \left(\int_{t_1}^{t_2} \iint M|(g_\varepsilon\gamma_\varepsilon)_{|\partial\Omega}|^2\frac{(v.n(x))^2 1_K(x)}{1 + |v|^2}dvd\sigma_x dt\right)^{1/2} \qquad (4.55)$$
$$\times \left(\iiint_\mathcal{A} M(1 + |v|^2)(1 + |v|^p)^2 dvd\sigma_x dt\right)^{1/2}.$$

Then, using (4.54), we obtain

$(\gamma_\varepsilon g_\varepsilon)_{|\partial\Omega}$ is equiintegrable in $L^1_{loc}(dtd\sigma_x; L^1(M(1+|v|^p)|v\cdot n(x)|dv))$

and thus relatively weakly compact by Dunford-Pettis theorem.

● Then we have to identify the limiting points of $(g_\varepsilon\gamma_\varepsilon)_{|\partial\Omega}$. From the limiting Boltzmann equation (4.50) we deduce that for all $\varphi \in C^1([0,T]\times\bar\Omega\times \mathbf{R}^3)$

$$\int_{t_1}^{t_2}\int_{\partial\Omega}\int M\varphi g_{|\partial\Omega}(v.n(x))dvd\sigma_x dt$$
$$= \int_{t_1}^{t_2}\int_\Omega\int M(v\cdot\nabla_x\varphi)gdvdxdt + 2\int_{t_1}^{t_2}\int_\Omega\int M\varphi q dvdxdt \tag{4.56}$$

On the other hand, by (2.31) we have,

$$\int_{t_1}^{t_2}\int_{\partial\Omega}\int M\varphi(g_\varepsilon\gamma_\varepsilon)_{|\partial\Omega}(v\cdot n(x))dvd\sigma_x\, dt$$
$$= \varepsilon\int_\Omega\int M\varphi g_\varepsilon\gamma_\varepsilon(t_1)dvdx - \varepsilon\int_\Omega\int M\varphi g_\varepsilon\gamma_\varepsilon(t_2)dvdx$$
$$+ \int_{t_1}^{t_2}\int_\Omega\int M(v\cdot\nabla_x\varphi)(g_\varepsilon\gamma_\varepsilon)(t,x,v)dvdxdt \tag{4.57}$$
$$+ \int_{t_1}^{t_2}\int_\Omega M\varphi(\varepsilon\partial_t + v\cdot\nabla_x)(g_\varepsilon\gamma_\varepsilon)dvdxdt$$

with

$$M(\varepsilon\partial_t + v\cdot\nabla_x)(g_\varepsilon\gamma_\varepsilon)$$
$$= \frac{1}{\varepsilon^2}\hat\gamma_\varepsilon Q(f_\varepsilon, f_\varepsilon)$$
$$= \frac{1}{\varepsilon^2}\iint \hat\gamma_\varepsilon(\sqrt{f'_\varepsilon f'_{\varepsilon*}} - \sqrt{f_\varepsilon f_{\varepsilon*}})^2 B(v-v_*,\sigma)dv_*d\sigma$$
$$+ \frac{2}{\varepsilon^2}\iint \hat\gamma_\varepsilon(\sqrt{f_\varepsilon}\sqrt{f_{\varepsilon*}} - 1)(\sqrt{f'_\varepsilon f'_{\varepsilon*}} - \sqrt{f_\varepsilon f_{\varepsilon*}})B(v-v_*,\sigma)dv_*d\sigma + 2M\hat\gamma_\varepsilon\hat q_\varepsilon$$
$$\stackrel{\text{def}}{=} 2M\hat\gamma_\varepsilon\hat q_\varepsilon + O(\varepsilon)_{L^1_{loc}(dtdx, L^1(dv))}$$

Then, taking limits in (4.57) and identifying with (4.56), we obtain

$(g_\varepsilon\gamma_\varepsilon)_{|\partial\Omega} \to g_{|\partial\Omega}$ weakly in $L^1_{loc}(dtd\sigma_x, L^1(M(1+|v|^p)|v\cdot n(x)|dv))$.

● It remains to establish the pointwise convergence of the trace of $g_\varepsilon\gamma_\varepsilon$. By definition of the trace $f_{\varepsilon|\partial\Omega}$, for any admissible renormalization Γ,

$$(\Gamma(f_\varepsilon))_{|\partial\Omega} = \Gamma(f_{\varepsilon|\partial\Omega}).$$

In particular,

$$(g_\varepsilon\gamma_\varepsilon)_{|\partial\Omega} = g_{\varepsilon|\partial\Omega}\gamma(1+\varepsilon g_{\varepsilon|\partial\Omega})$$

Then, by (4.52),

$\varepsilon(g_\varepsilon\gamma_\varepsilon)_{|\partial\Omega} = \varepsilon g_{\varepsilon|\partial\Omega}\gamma(1+\varepsilon g_{\varepsilon|\partial\Omega}) \to 0$ almost everywhere on $\mathbf{R}^+\times\partial\Omega\times\mathbf{R}^3$,

which concludes the proof.   □

4.4.2 A Priori Estimates Coming from the Boundary

Let us first recall from Chapter 3 that the uniform bound on the scaled Darrozès-Guiraud information

$$\frac{\alpha_\varepsilon}{\varepsilon} \int_0^t \int_{\partial\Omega} E(f_\varepsilon|M)(s,x)d\sigma_x ds \leq H(f_{\varepsilon,in}|M). \tag{4.58}$$

allows to control the renormalized trace variation $\hat{\eta}_\varepsilon$ defined by

$$\hat{\eta}_\varepsilon = \frac{2}{\varepsilon}\sqrt{\frac{\alpha_\varepsilon}{\varepsilon}} 1_{\Sigma_+} \left(\sqrt{\frac{f_\varepsilon}{M}} - \sqrt{\left\langle \frac{f_\varepsilon}{M} \right\rangle_{\partial\Omega}} \right) \tag{4.59}$$

where $\langle . \rangle_{\partial\Omega}$ denotes as previously the average (with respect to v) of any quantity defined on the boundary

$$\langle g \rangle_{\partial\Omega} = \sqrt{2\pi} \int M g_{|\partial\Omega}(v \cdot n)_+ dv.$$

We have indeed

$$\|\hat{\eta}_\varepsilon\|_{L^2(dtd\sigma_x M(v \cdot n)_+ dv)} \leq C_{in}. \tag{4.60}$$

In the case where

$$\frac{\alpha_\varepsilon}{\sqrt{2\pi}\varepsilon} \to \lambda \in]0,+\infty],$$

this estimate, combined with the convergence studied in Proposition 4.4.1, allows to identify the limiting trace.

Proposition 4.4.2 *With the same notations and assumptions as in Theorem 4.1.6, define*

$$\Delta_\varepsilon = 1_{\Sigma_+} \left(g_{\varepsilon|\partial\Omega} - \langle g_\varepsilon \rangle_{\partial\Omega} \right) . \tag{4.61}$$

Assume that

$$\frac{\alpha_\varepsilon}{\varepsilon} \to \sqrt{2\pi}\lambda \in]0,+\infty].$$

Then,

$$\gamma_{\varepsilon|\partial\Omega}\Delta_\varepsilon \rightharpoonup 1_{\Sigma_+} \left(g_{|\partial\Omega} - \langle g \rangle_{\partial\Omega} \right)$$

weakly in $L^1_{loc}(dt, L^1(M|v \cdot n(x)|d\sigma_x dv))$.

Proof. In order to establish that result, we have to identify the weak limit using both the convergence coming from the inside and the convergence coming from the boundary.

- Start from the identity

$$\Delta_\varepsilon = \sqrt{\frac{\varepsilon}{\alpha_\varepsilon}}\hat{\eta}_\varepsilon\sqrt{\langle 1 + \varepsilon g_\varepsilon \rangle_{\partial\Omega}} + \frac{\varepsilon^2}{4\alpha_\varepsilon}\hat{\eta}_\varepsilon^2. \tag{4.62}$$

By Proposition 4.4.1, $\gamma_{\varepsilon|\partial\Omega}$ is bounded and converges pointwise to 1 and

$$\gamma_{\varepsilon|\partial\Omega}\sqrt{1+\varepsilon g_{\varepsilon|\partial\Omega}} \to 1 \text{ in } L^2_{loc}(dtd\sigma_x; L^2(M(1+|v|)^p(v\cdot n(x))_+ dv))$$

Therefore, by the L^2 estimate on $\hat{\eta}_\varepsilon$ and the Product Limit Theorem,

$$\mathbf{1}_{\Sigma_+}\gamma_{\varepsilon|\partial\Omega}\sqrt{\langle 1+\varepsilon g_\varepsilon\rangle_{\partial\Omega}} = \mathbf{1}_{\Sigma_+}\gamma_{\varepsilon|\partial\Omega}\left(\sqrt{1+\varepsilon g_\varepsilon} - \frac{\varepsilon}{2}\sqrt{\frac{\varepsilon}{\alpha_\varepsilon}}\hat{\eta}_\varepsilon\right) \tag{4.63}$$
$$\to \mathbf{1}_{\Sigma_+} \text{ in } L^2_{loc}(dtd\sigma_x, L^2(M(v\cdot n)_+ dv))$$

Using again the Product Limit theorem and the L^2 estimate on $\hat{\eta}_\varepsilon$ we then deduce that, up to extraction of a subsequence,

$$\gamma_{\varepsilon|\partial\Omega}\Delta_\varepsilon \text{ converges weakly in } L^1_{loc}(dtd\sigma_x, L^1(M(v\cdot n(x))_+ dv)). \tag{4.64}$$

On the other hand, using the definition (4.61) of Δ_ε and the definition of the trace $g_{\varepsilon|\partial\Omega}$, we get

$$\gamma_{\varepsilon|\partial\Omega}\Delta_\varepsilon = \mathbf{1}_{\Sigma_+}\left((g_\varepsilon\gamma_\varepsilon)_{|\partial\Omega} - \gamma_{\varepsilon|\partial\Omega}\langle g_\varepsilon\rangle_{\partial\Omega}\right).$$

From the convergence statement (4.52) obtained from the inside and the convergence (4.64) coming from the boundary term in the entropy inequality, we deduce that

$$\gamma_{\varepsilon|\partial\Omega}\langle g_\varepsilon\rangle_{\partial\Omega} \text{ converges weakly in } L^1_{loc}(dtd\sigma_x, L^1(M(v\cdot n(x))_+ dv)). \tag{4.65}$$

- It remains then to identity the limit of this averaged part. We first prove that it is of the form $\mathbf{1}_{\Sigma_+}\rho$ for some $\rho \in L^1_{loc}(dtd\sigma_x)$.
  Indeed define

$$\rho_\varepsilon = \langle g_\varepsilon\rangle_{\partial\Omega}\left(\gamma_{\varepsilon|\Sigma_+} + L(\gamma_{\varepsilon|\Sigma_+})\right)$$

where L denotes - as in Chapter 2 - the local reflection operator. Hence, by (4.65), there exists $\rho \in L^1_{loc}(dtd\sigma_x, L^1(M|v\cdot n(x)|dv))$ such that

$$\rho_\varepsilon \to \rho \quad \text{and} \quad \mathbf{1}_{\Sigma_+}\gamma_{\varepsilon|\partial\Omega}\langle g_\varepsilon\rangle_{\partial\Omega} = \mathbf{1}_{\Sigma_+}\rho_\varepsilon \to \mathbf{1}_{\Sigma_+}\rho$$

weakly in $L^1_{loc}(dtd\sigma_x, L^1(M|v\cdot n(x)|dv))$.
  Because $\langle g_\varepsilon\rangle_{\partial\Omega}$ depends only on t and x, and

$$\left(\gamma_{\varepsilon|\Sigma_+} + L(\gamma_{\varepsilon|\Sigma_+})\right)$$

is bounded and converges a.e. to 1 on $\mathbf{R}^+\times\partial\Omega\times\mathbf{R}^3$, we deduce from a result of variables separating, which is actually a variant of the Product Limit theorem (see Proposition A.2 in Appendix A) that

$$\rho \equiv \rho(t, x).$$

Gathering all results together, we thus have

$$\gamma_{\varepsilon|\partial\Omega}\Delta_\varepsilon \to \mathbf{1}_{\Sigma_+}(g_{|\partial\Omega} - \rho) \text{ in } L^1_{loc}(dtd\sigma_x, L^1(M(v\cdot n(x))_+ dv)).$$

We also point out that there is no possible mass going to infinity in the v-variable since

$$\Delta_\varepsilon = \sqrt{\frac{\varepsilon}{\alpha_\varepsilon}}\hat{\eta}_\varepsilon + \sqrt{\frac{\varepsilon}{\alpha_\varepsilon}}\hat{\eta}_\varepsilon \left(\sqrt{\langle 1 + \varepsilon g_\varepsilon\rangle_{\partial\Omega}} - 1\right) + \frac{\varepsilon^2}{4\alpha_\varepsilon}\hat{\eta}_\varepsilon^2$$

is the sum of a term bounded in $L^2_{loc}(dt, L^2(M(v \cdot n(x))_+ dv d\sigma_x))$ and terms going strongly to 0 in $L^1_{loc}(dt, L^1(M(v \cdot n(x))_+ d\sigma_x dv))$. Then, because $\langle\Delta_\varepsilon\rangle_{\partial\Omega} = 0$,

$$\rho = \langle g\rangle_{\partial\Omega}, \tag{4.66}$$

which concludes the proof. □

4.4.3 The Limiting Boundary Conditions

Using the results of the two previous paragraphs, we can now pass to the limit in the boundary conditions. We start with the Dirichlet boundary condition, which turns out to be easier because we recover it directly.

The Limiting Dirichlet Boundary Condition

If we assume that $\alpha_\varepsilon/\varepsilon \to +\infty$, then the uniform L^2 estimate on $\hat{\eta}_\varepsilon$ shows that

$$\gamma_{\varepsilon|\partial\Omega}\Delta_\varepsilon = \sqrt{\frac{\varepsilon}{\alpha_\varepsilon}}\hat{\eta}_\varepsilon\sqrt{\langle 1 + \varepsilon g_\varepsilon\rangle_{\partial\Omega}}\gamma_{\varepsilon|\partial\Omega} + \frac{\varepsilon^2}{4\alpha_\varepsilon}\hat{\eta}_\varepsilon^2\gamma_{\varepsilon|\partial\Omega}$$

converges to 0 strongly in $L^1_{loc}(dt d\sigma_x, L^1(M(v \cdot n(x))_+ dv))$.
    Proposition 4.4.2 implies then that

$$\mathbf{1}_{\Sigma_+}(g_{|\partial\Omega} - \langle g\rangle_{\partial\Omega}) = 0.$$

Finally, by (4.51), we have necessarily

$$u_{|\partial\Omega} = 0 \text{ and } \theta_{|\partial\Omega} = 0.$$

The Limiting Navier Boundary Condition

• The condition of zero mass flux

$$n \cdot u_{|\partial\Omega} = 0$$

(which corresponds to the incompressibility constraint inside the domain Ω) is obtained in a very simple way, using only the weak compactness statement and the pointwise convergence established in Proposition 4.4.1.

The renormalized form (2.32) of Maxwell's boundary condition reads indeed

$$
(g_\varepsilon \gamma_\varepsilon)_{|\Sigma_-}
= \left((1 - \alpha_\varepsilon) L(g_{\varepsilon | \Sigma_+}) + \alpha_\varepsilon \langle g_\varepsilon \rangle_{\partial\Omega} \right) \gamma \left(1 + \varepsilon \left((1 - \alpha_\varepsilon) L(g_{\varepsilon | \Sigma_+}) + \alpha_\varepsilon \langle g_\varepsilon \rangle_{\partial\Omega} \right) \right).
$$

Then, assuming that $\alpha_\varepsilon = O(\varepsilon)$ and passing to the limit in the sense of distributions, or, more precisely, weakly in $L^1_{loc}(dt d\sigma_x, L^1(M|v \cdot n(x)|dv))$, we get that

$$
g_{|\Sigma_-} = L g_{|\Sigma_+},
$$

from which we deduce easily that $n \cdot u_{|\partial\Omega} = 0$.

• The main difficulty is (as for the inside) to take limits in the conservation laws written in weak form. Start from the weak forms of (4.24) and (4.25), namely

$$
\int_\Omega w \cdot \int M \mathbf{1}_{|v|^2 \leq K_\varepsilon} v g_\varepsilon \gamma_\varepsilon(t_2, x, v) dv dx - \int_\Omega w \cdot \int M \mathbf{1}_{|v|^2 \leq K_\varepsilon} v g_\varepsilon \gamma_\varepsilon(t_1, x, v) dv dx
$$

$$
- \int_{t_1}^{t_2} \int_\Omega \nabla_x w : F_\varepsilon(\Phi)(t, x) dx dt - \int_{t_1}^{t_2} \int_\Omega w \cdot D_\varepsilon(v)(t, x) dx dt
$$

$$
= -\frac{1}{\varepsilon} \int_{t_1}^{t_2} \int_{\partial\Omega} w \cdot \int M \mathbf{1}_{|v|^2 \leq K_\varepsilon} v g_\varepsilon \gamma_\varepsilon(t, x, v)(v \cdot n(x)) dv d\sigma_x dt,
$$

and

$$
\int_\Omega \varphi \int M \mathbf{1}_{|v|^2 \leq K_\varepsilon} \left(\frac{|v|^2}{5} - 1 \right) g_\varepsilon \gamma_\varepsilon(t_2, x, v) dv dx
$$

$$
- \int_\Omega \varphi \int M \mathbf{1}_{|v|^2 \leq K_\varepsilon} \left(\frac{|v|^2}{5} - 1 \right) g_\varepsilon \gamma_\varepsilon(t_1, x, v) dv dx
$$

$$
- \frac{2}{5} \int_{t_1}^{t_2} \int_\Omega \nabla_x \varphi \cdot F_\varepsilon(\Psi)(t, x) dx dt - \int_{t_1}^{t_2} \int_\Omega \varphi D_\varepsilon \left(\frac{|v|^2}{5} - 1 \right)(t, x) dx dt
$$

$$
= -\frac{1}{\varepsilon} \int_{t_1}^{t_2} \int_{\partial\Omega} \varphi \int M \mathbf{1}_{|v|^2 \leq K_\varepsilon} \left(\frac{|v|^2}{5} - 1 \right) g_\varepsilon \gamma_\varepsilon(t, x, v)(v \cdot n(x)) dv d\sigma_x dt,
$$

for all divergence free vector field $w \in C_c^\infty(\bar{\Omega})$ with $n \cdot w_{|\partial\Omega} = 0$, and all $\varphi \in C_c^\infty(\bar{\Omega})$. Then taking limits in the flux terms and conservation defects as in the previous section leads to

$$
\int_\Omega w \cdot u(t_2, x) dx - \int_\Omega w \cdot u(t_1, x) dx + \lim_{\varepsilon \to 0} B_\varepsilon(w \cdot v)
$$

$$
= \int_{t_1}^{t_2} \int_\Omega \nabla_x w : \left(u \otimes u - \mu(\nabla_x u + (\nabla_x u)^T) \right)(t, x) dx dt
$$

and $\int_\Omega \varphi \theta(t_2, x) dx - \int_\Omega \varphi \theta(t_1, x) dx + \lim_{\varepsilon \to 0} B_\varepsilon \left(\varphi \left(\frac{|v|^2}{5} - 1 \right) \right)$

$$
= \int_{t_1}^{t_2} \int_\Omega \nabla_x \varphi \cdot (u\theta - \kappa \nabla_x \theta)(t, x) dx dt
$$

(4.67)

where

$$B_\varepsilon(\zeta) = \frac{1}{\varepsilon} \int_{t_1}^{t_2} \int_{\partial\Omega} \int M \mathbf{1}_{|v|^2 \leq K_\varepsilon} \zeta g_\varepsilon \gamma_\varepsilon(t, x, v)(v \cdot n(x)) dv d\sigma_x dt. \qquad (4.68)$$

Next, in order to obtain the limiting boundary conditions, we have to compute the limits of $B_\varepsilon(w \cdot v)$ and $B_\varepsilon\left(\varphi\left(|v|^2/5 - 1\right)\right)$. We start by expressing $B_\varepsilon(\zeta)$ in terms of the controlled quantities. The crucial remark here is that $L\zeta = \zeta$ a.e. on $\partial\Omega \times \mathbf{R}^3$ (using the zero mass flux condition $n \cdot w_{|\partial\Omega} = 0$). We therefore have

$$\int M \mathbf{1}_{|v|^2 \leq K_\varepsilon} \zeta g_\varepsilon \gamma_\varepsilon(v \cdot n(x)) dv d\sigma_x dt$$
$$= \int M \mathbf{1}_{|v|^2 \leq K_\varepsilon} \zeta (\mathbf{1}_{\Sigma_+} g_\varepsilon \gamma_\varepsilon - L(\mathbf{1}_{\Sigma_-} g_\varepsilon \gamma_\varepsilon))(v \cdot n(x))_+ dv$$

Using Taylor's formula, we get

$$g_\varepsilon \gamma_\varepsilon - L(\mathbf{1}_{\Sigma_-} g_\varepsilon \gamma_\varepsilon)$$

$$= g_\varepsilon \gamma(1 + \varepsilon g_\varepsilon) - \left((1 - \alpha_\varepsilon)g_\varepsilon + \alpha_\varepsilon \langle g_\varepsilon \rangle_{\partial\Omega}\right) \gamma\left(1 + \varepsilon\left((1 - \alpha_\varepsilon)g_\varepsilon + \alpha_\varepsilon \langle g_\varepsilon \rangle_{\partial\Omega}\right)\right)$$

$$= \alpha_\varepsilon(g_\varepsilon - \langle g_\varepsilon \rangle_{\partial\Omega}) \int_0^1 \hat\gamma_{\varepsilon,\tau} d\tau$$

where we denote

$$\hat\gamma_{\varepsilon,\tau} = \hat\gamma\left((1 - \tau\alpha_\varepsilon)(1 + \varepsilon g_\varepsilon) + \tau\alpha_\varepsilon \langle 1 + \varepsilon g_\varepsilon \rangle_{\partial\Omega}\right) \text{ with } \hat\gamma(z) = \gamma(z) + (z - 1)\gamma'(z).$$

Therefore

$$B_\varepsilon(\zeta) = \frac{\alpha_\varepsilon}{\sqrt{2\pi\varepsilon}} \int_{t_1}^{t_2} \int_{\partial\Omega} \int_0^1 \left\langle \mathbf{1}_{|v|^2 \leq K_\varepsilon} \zeta \Delta_\varepsilon \hat\gamma_{\varepsilon,\tau} \right\rangle_{\partial\Omega}(t, x) d\tau d\sigma_x dt \qquad (4.69)$$

As we have assumed that $\alpha_\varepsilon/\varepsilon \to \sqrt{2\pi}\lambda < +\infty$, for ε sufficiently small, we further have

$$\hat\gamma_{\varepsilon,\tau} \leq \frac{C}{1 + (1 - \tau\alpha_\varepsilon)(1 + \varepsilon g_\varepsilon) + \tau\alpha_\varepsilon \langle 1 + \varepsilon g_\varepsilon \rangle_{\partial\Omega}} \leq \frac{2C}{3 + \varepsilon g_\varepsilon}.$$

- If $\lambda = 0$, by (4.62), the decomposition in (4.63) together with Proposition 4.4.1 and the L^2 estimate on $\hat\eta_\varepsilon$ shows that

$$\sqrt{\frac{\alpha_\varepsilon}{\varepsilon}} \hat\gamma_{\varepsilon,\tau} \sqrt{\langle 1 + \varepsilon g_\varepsilon \rangle_{\partial\Omega}} = O\left(\sqrt{\frac{\alpha_\varepsilon}{\varepsilon}}\right)_{L^2_{loc}(dtd\sigma_x, L^2((1+|v|)^p(v \cdot n(x))_+ M dv))}$$
$$+ O(\varepsilon)_{L^2_{loc}(dtd\sigma_x, L^2((v \cdot n(x))_+ M dv))}$$

We then have

$$\frac{\alpha_\varepsilon}{\varepsilon}\Delta_\varepsilon\hat{\gamma}_{\varepsilon,\tau} = O\left(\sqrt{\frac{\alpha_\varepsilon}{\varepsilon}}\right)_{L^1_{loc}(dtd\sigma_x,L^1((1+|v|)^p(v\cdot n(x))_+Mdv))} \\ +O(\varepsilon)_{L^1_{loc}(dtd\sigma_x,L^1((v\cdot n(x))_+Mdv))}$$

from which we deduce, using the L^∞ bound on $\zeta\mathbf{1}_{|v|^2\leq K_\varepsilon}$, that

$$B_\varepsilon(\zeta) \to 0.$$

- If $\lambda \in]0,+\infty[$, using both the identity (4.62) and the decomposition in (4.63), we get

$$\Delta_\varepsilon\hat{\gamma}_{\varepsilon,\tau} = \sqrt{\frac{\varepsilon}{\alpha_\varepsilon}}\hat{\eta}_\varepsilon\hat{\gamma}_{\varepsilon,\tau}\sqrt{1+\varepsilon g_\varepsilon} + O(\varepsilon)_{L^1_{loc}(dtd\sigma_x,L^1((v\cdot n(x))_+Mdv))}.$$

The first term in the right-hand side is uniformly bounded in $L^2_{loc}(dtd\sigma_x, L^2((v\cdot n(x))_+Mdv))$, and we can identify its weak limit using Proposition 4.4.2 :

$$\sqrt{\frac{\varepsilon}{\alpha_\varepsilon}}\hat{\eta}_\varepsilon\hat{\gamma}_{\varepsilon,\tau}\sqrt{1+\varepsilon g_\varepsilon} \to \mathbf{1}_{\Sigma_+}(g_{|\partial\Omega} - \langle g\rangle_{\partial\Omega}).$$

We then conclude, using again the L^∞ bound on $\zeta\mathbf{1}_{|v|^2\leq K_\varepsilon}$, that

$$B_\varepsilon(\zeta) \to \lambda\int_{t_1}^{t_2}\int_{\partial\Omega}\left\langle\zeta(g - \langle g\rangle_{\partial\Omega})\right\rangle_{\partial\Omega}(t,x)d\sigma_x dt.$$

Because u and w satisfy the zero mass flux condition, we can use (4.51) to compute both limits in terms of $u_{|\partial\Omega}$ and $\theta_{|\partial\Omega}$. We then obtain the weak form of the Navier-Stokes Fourier system with Navier boundary conditions :

$$\int_\Omega w\cdot u(t_2,x)dx - \int_\Omega w\cdot u(t_1,x)dx + \lambda\int_{t_1}^{t_2}\int_{\partial\Omega}w\cdot u(t,x)d\sigma_x dt$$
$$= \int_{t_1}^{t_2}\int_\Omega\nabla_x w : \left(u\otimes u - \mu(\nabla_x u + (\nabla_x u)^T)\right)(t,x)dxdt$$

and $$\int_\Omega\varphi\theta(t_2,x)dx - \int_\Omega\varphi\theta(t_1,x)dx + \frac{4}{5}\lambda\int_{t_1}^{t_2}\int_{\partial\Omega}\varphi\theta(t,x)d\sigma_x dt$$
$$= \int_{t_1}^{t_2}\int_\Omega\nabla_x\varphi\cdot(u\theta - \kappa\nabla_x\theta)(t,x)dxdt,$$

which concludes the proof.   □

5

The Incompressible Euler Limit

The aim of this chapter is to describe the state of the art about the incompressible Euler limit of the Boltzmann equation, which is not so complete as the incompresible Navier-Stokes limit presented in the previous chapter.

Due to the lack of regularity estimates for inviscid incompressible models, the convergence results describing the incompressible Euler asymptotics of the Boltzmann equation require additional regularity assumptions on the solution to the target equations.

Furthermore, the relative entropy method leading to these stability results controls the convergence in a very strong sense, which imposes additional conditions either on the solution to the asymptotic equations ("well-prepared initial data"), or on the solutions to the scaled Boltzmann equation (namely some additional non uniform a priori estimates giving in particular the local conservation of momentum and energy).

5.1 Convergence Result : From the Boltzmann Equation to the Incompressible Euler System

5.1.1 Mathematical Theories for the Incompressible Euler Equations

Before giving the precise mathematical statement of the convergence result, let us first recall some basic facts about the limiting system.

The incompressible Euler equations govern the velocity field $u \equiv u(t,x)$ of an inviscid incompressible homogeneous fluid. They are

$$\partial_t u + (u \cdot \nabla_x)u + \nabla_x p = 0, \quad \nabla_x \cdot u = 0. \tag{5.1}$$

The first equality in (5.1) states that the fluid preserves the volume, and is referred to as the *incompressibility condition*; the second equality expresses Newton's law of dynamics for an infinitesimal volume of fluid.

L. Saint-Raymond, *Hydrodynamic Limits of the Boltzmann Equation*,
Lecture Notes in Mathematics 1971, DOI: 10.1007/978-3-540-92847-8_5,
© Springer-Verlag Berlin Heidelberg 2009

Note that (5.1) can be formally obtained from the incompressible Navier-Stokes equations (4.1) in the vanishing viscosity limit, so taking $\mu = 0$. In particular the fundamental energy estimate becomes:

Proposition 5.1.1 *Let $u \equiv u(t, x)$ be a solution of the incompressible Euler equations (5.1) on $\mathbf{R}^+ \times \Omega$ that is sufficiently smooth (which belongs for instance to $C(\mathbf{R}^+, H^1(\Omega))$) and satisfies the zero mass flux condition*

$$n \cdot u_{|\partial\Omega} = 0,$$

where n denotes as previously the outward unit normal to $\partial\Omega$. Then the following energy estimate holds :

$$\|u(t)\|_{L^2(\Omega)}^2 = \|u_{in}\|_{L^2(\Omega)}^2 \,. \tag{5.2}$$

The crucial difference with the energy estimate (4.3) obtained for the incompressible Navier-Stokes equations is that (5.2) does not provide any spatial regularity on the velocity field u. This lack of a priori estimates induces many difficulties when dealing with the Cauchy problem for (5.1).

Local Smooth Solutions

The first result concerning the Cauchy problem (5.1) has been obtained by Lichtenstein [71] in the framework of $C^{1,1}$ initial data.

It has been later improved by many authors. In particular, Beale, Kato and Majda [10] have established the following sharp persistence criterion involving the vorticity :

Theorem 5.1.2 *Let $u_{in} \in H^s(\Omega)$ for $s > 5/2$ be a divergence free vector field. Then there exists a unique maximal solution*

$$u \in L^\infty_{loc}([0, t^*), H^s(\Omega))$$

to the incompressible Euler equations (5.1) with initial data u_{in} and supplemented by the boundary condition

$$n \cdot u_{|\partial\Omega} = 0, \tag{5.3}$$

which satisfies in addition

$$\int_0^{t^*} \|\mathrm{rot}\,_x u(t)\|_{L^\infty(\Omega)} dt = +\infty \,. \tag{5.4}$$

The local existence of such a smooth solution is obtained using a standard approximation scheme, and some a priori bounds based on trilinear estimates of the following type :

$$\langle w, (u \cdot \nabla_x) w \rangle_{H^s(\mathbf{R}^3)} \leq C \|w\|_{H^s(\mathbf{R}^3)}^2 \|u\|_{H^s(\mathbf{R}^3)}$$

for divergence free vector fields. That determines in particular the critical regularity $s_0 = 5/2$ required on the initial data.

The extension of such trilinear estimates to the case of domains with boundaries requires elliptic estimates and composition theorems which can be found for instance in the paper [20] by Brézis and Bourguignon.

The study of persistence is based on some precised energy inequalities very similar to (4.8) which show that the propagation of the H^s norm on $[0, t]$ is controlled by the quantity

$$\|\nabla_x u + (\nabla_x u)^T\|_{L^1([0,t],L^\infty(\Omega))}.$$

A refined study using crucially the fact that u is a divergence free vector field, shows that the propagation of the H^s norm on $[0, t]$ is actually controlled by the vorticity

$$\|\mathrm{rot}\,_x u\|_{L^1([0,t],L^\infty(\Omega))}$$

(we refer to [10] for the proof of the refined blow-up condition (5.4)).

Remark 5.1.3 *Note that such a local existence result can be improved in two space dimensions. First of all the critical regularity becomes $s_0 = 2$. Furthermore the vorticity $\omega = \partial_2 u_1 - \partial_1 u_2$ satisfies the transport equation*

$$\partial_t \omega + u \cdot \nabla_x \omega = 0, \qquad (5.5)$$

so that, assuming that the initial vorticity ω_{in} is bounded in $L^\infty(\Omega)$, Theorem 5.1.2 provides a global existence result.

Strong-Weak Stability

Blow up criteria are therefore based on propagation results, which are obtained as energy estimates :

Proposition 5.1.4 *Let $u \in L^\infty([0, t^*], L^2(\Omega)) \cap C([0, t^*], w - L^2(\Omega))$ be a weak solution to the incompressible Euler equations (5.1)(5.3) with initial data u_{in} satisfying the energy inequality*

$$\forall t \in [0, t^*], \quad \|u(t)\|^2_{L^2(\Omega)} \leq \|u_{in}\|^2_{L^2(\Omega)}.$$

Let $\tilde{u} \in C([0, t^*], L^2(\Omega))$ be a strong solution to (5.1)(5.3) with initial data $\tilde{u}_{in}$ such that

$$\int_0^{t^*} \|\nabla_x \tilde{u} + (\nabla_x \tilde{u})^T(t)\|_{L^\infty(\Omega)} dt < +\infty.$$

Then the following stability inequality holds for all $t \leq t^*$,

$$\|u(t) - \tilde{u}(t)\|^2_{L^2(\Omega)} \leq \|u_{in} - \tilde{u}_{in}\|^2_{L^2(\Omega)} \exp\left(\int_0^t \|(\nabla \tilde{u} + (\nabla \tilde{u})^T)(s)\|_{L^\infty(\Omega)} ds\right).$$
$$(5.6)$$

In particular, $u = \tilde{u}$ on $[0, t^*] \times \Omega$ if $u_{in} = \tilde{u}_{in}$.

Proof. In order to obtain the stability estimate, we start by writing the equation satisfied by $w = u - \tilde{u}$:

$$\partial_t w + u \cdot \nabla w = -\nabla p - w \cdot \nabla \tilde{u},$$

from which we deduce the formal L^2 estimate :

$$\|w(t)\|^2_{L^2(\Omega)} - \|w_{in}\|^2_{L^2(\Omega)}$$
$$\leq -2 \int_0^t \int (u \cdot \nabla) w \cdot w(s, x) dx ds - 2 \int_0^t \int (w \cdot \nabla) \tilde{u} \cdot w(s, x) dx ds$$
$$\leq \int_0^t \int \|(\nabla \tilde{u} + (\nabla \tilde{u})^T)\|_{L^\infty(\Omega)} \|w\|^2_{L^2(\Omega)}(s) ds$$

because u and w are divergence free vector fields and satisfy the zero mass flux condition (5.3). (In order to justify this formal computation one should of course proceed by approximation). We then conclude using Gronwall's lemma.   □

Weak Solutions

In two space dimensions, because of the special structure of the vorticity equation (5.5), one has a global unique smooth solution to (5.1) as soon as $\omega_{in} \in L^\infty(\Omega)$.

Furthermore one can build global weak solutions requiring much less regularity on the initial data : see for instance the results by Yudovitch [110] in the case

$$u_{in} \in L^2(\mathbf{R}^2), \quad \omega_{in} \in L^p(\mathbf{R}^2) \text{ with } p > 1,$$

and those by Delort [41] in the case of vortex sheets

$$u_{in} \in L^2(\mathbf{R}^2), \quad \omega_{in} \in \mathcal{M}^+(\mathbf{R}^2).$$

Nevertheless, we have no uniqueness for these solutions. In a recent paper [40], De Lellis and Székelyhidi have indeed studied various admissibility criteria that could be imposed on weak solutions of Euler, and they have shown that none of these criteria implies uniqueness for general L^2 initial data. More precisely, they have proved the following

Proposition 5.1.5 *There exist bounded and compactly supported divergence-free vector fields u_{in} for which there are*

- *infinitely many weak solutions of (5.1) satisfying both the strong energy equality*

$$\|u(t)\|^2_{L^2} = \|u(s)\|^2_{L^2} \text{ for every pair } (s, t) \text{ with } s < t, \qquad (5.7)$$

and the local energy equality

$$\partial_t |u|^2 + \nabla_x \cdot (u(|u|^2 + 2p)) = 0; \qquad (5.8)$$

- *weak solutions of (5.1) satisfying the strong energy inequality*

$$\|u(t)\|_{L^2}^2 \le \|u(s)\|_{L^2}^2 \text{ for every pair } (s,t) \text{ with } s < t, \qquad (5.9)$$

  but not the energy equality (5.7);
- *weak solutions of (5.1) satisfying the weak energy inequality*

$$\|u(t)\|_{L^2}^2 \le \|u(s)\|_{L^2}^2 \text{ for almost every pair } (s,t) \text{ with } s < t, \qquad (5.10)$$

  but not the strong energy inequality (5.9).

Their examples display a very wild behavior, such as dissipation of the energy and amplitude of high-frequency oscillations.

Dissipative Solutions

In three space dimensions the question of the global continuation of solutions (even in weak sense) remains completely open. As an alternative, starting from (5.6), Lions [74] has proposed the following very weak notion of solution :

Definition 5.1.6 *A dissipative solution on $[0, t^*)$ to (5.1)(5.3) with initial data u_{in} is a function*

$$u \in L^\infty([0,t^*), L^2(\Omega)) \cap C([0,t^*), w - L^2(\Omega))$$

satisfying $\nabla \cdot u = 0$ and $u_{|t=0} = u_{in}$ in the sense of distributions and such that

$$\|u(t) - \tilde{u}(t)\|_{L^2(\Omega)}^2 \le \|u_{in} - \tilde{u}_{in}\|_{L^2(\Omega)}^2 \exp\left(\int_0^t \|(\nabla\tilde{u} + (\nabla\tilde{u})^T)(s)\|_{L^\infty(\Omega)} ds\right)$$
$$+ \int_0^t \int A(\tilde{u}) \cdot (\tilde{u} - u)(s,x) dx \exp\left(\int_s^t \|(\nabla\tilde{u} + (\nabla\tilde{u})^T)(\tau)\|_{L^\infty(\Omega)} d\tau\right) \qquad (5.11)$$

for all $t \in [0, t^*)$ and all test functions $\tilde{u} \in C([0, t^*] \times \Omega)$ such that

$$\nabla \cdot \tilde{u} = 0, \quad n \cdot \tilde{u}_{|\partial\Omega} = 0,$$

$$(\nabla\tilde{u} + (\nabla\tilde{u})^T) \in L^1([0, t^*], L^\infty(\Omega)), \qquad (5.12)$$

$$A(\tilde{u}) = \partial_t\tilde{u} + (\tilde{u} \cdot \nabla)\tilde{u} \in L^1([0, t^*], L^2(\Omega)).$$

Such solutions always exist, they are not known to be weak solutions of (5.1) in conservative form, but they coincide with the unique smooth solution with same initial data as long as the latter does exist. Indeed the stability principle stated in Proposition 5.1.4 can be extended as follows

Proposition 5.1.7 *Let $u \in L^\infty([0, t^*], L^2(\Omega)) \cap C([0, t^*), w - L^2(\Omega))$ be any dissipative solution to the incompressible Euler equations (5.1)(5.3) with initial data u_{in}, and $\tilde{u} \in C([0, t^*], L^2(\Omega))$ be a strong solution to (5.1)(5.3) with initial data $\tilde{u}_{in}$ such that*

$$\int_0^{t^*} \|\nabla_x \tilde{u} + (\nabla_x \tilde{u})^T (t)\|_{L^\infty(\Omega)} dt < +\infty,$$

$$\int_0^{t^*} \|\partial_t \tilde{u} + (\tilde{u} \cdot \nabla)\tilde{u}\|_{L^2(\Omega)} dt < +\infty.$$

Then the following stability inequality holds for all $t \leq t^*$,

$$\|u(t) - \tilde{u}(t)\|_{L^2(\Omega)}^2 \leq \|u_{in} - \tilde{u}_{in}\|_{L^2(\Omega)}^2 \exp\left(\int_0^t \|(\nabla \tilde{u} + (\nabla \tilde{u})^T)(s)\|_{L^\infty(\Omega)} ds\right).$$

(5.13)

In particular, $u = \tilde{u}$ on $[0, t^*] \times \Omega$ if $u_{in} = \tilde{u}_{in}$.

Note that the notion of dissipative solution has been introduced especially to investigate the inviscid limit of the incompressible Navier-Stokes equations, and then used to study various asymptotics such as the gyrokinetic or quasineutral limits of the Vlasov-Poisson equation [17].

5.1.2 Analogies with the Scaled Boltzmann Equation

As mentioned in Chapter 4 in the framework of the Navier-Stokes limit, a possibility to describe the incompressible hydrodynamic limits of the Boltzmann equation is to get some counterpart at kinetic level of the strong-weak uniqueness principle stated in Proposition 5.1.7.

The functional which measures the stability for the scaled Boltzmann equation is obtained naturally from the relative entropy

$$H(f_\varepsilon|M) = \iint \left(f_\varepsilon \log \frac{f_\varepsilon}{M} - f_\varepsilon + M \right) dv dx$$

which is a nonnegative Lyapunov functional for the Boltzmann equation (see Boltzmann's H theorem in Chapter 2), and controls the size of the fluctuation in incompressible regimes (see Lemmas 3.1.2 and 3.1.3 in Chapter 3). The idea of using the notion of relative entropy for this kind of problems comes actually from the notion of entropic convergence developed by C. Bardos, F. Golse and C.D. Levermore in [5], and on the other hand from Yau's elegant derivation of the hydrodynamic limit of the Ginzburg-Landau lattice model [108].

The *modulated entropy* is then defined for each test field $(R, U, T) \in C_c^\infty(\mathbf{R}_+ \times \Omega)$ by

$$H(f_\varepsilon|\mathcal{M}_{(R,U,T)}) \overset{\text{def}}{=} \iint \left(f_\varepsilon \log \frac{f_\varepsilon}{\mathcal{M}_{(R,U,T)}} - f_\varepsilon + \mathcal{M}_{(R,U,T)} \right) dv dx$$

$$= H(f_\varepsilon|M) + \iint \left(-\log \frac{R}{T^{3/2}} + \frac{|v - U|^2}{2T} - \frac{|v|^2}{2} \right) f_\varepsilon dv dx \quad (5.14)$$

$$- \iint (f_\varepsilon - \mathcal{M}_{(R,U,T)}) dv dx$$

Our goal here will be therefore to establish a stability inequality on the modulated entropy of the same type as (5.1.7), in the scaling leading to the incompressible Euler limit, i.e. for

$$R = \exp(\varepsilon\rho),$$
$$U = \varepsilon u,$$
$$T = \exp(\varepsilon\theta).$$

Note that, because of a poor understanding of the limiting system, we are not able at the present time to develop alternative strategies, leading for instance to weak convergence results, nor to consider spatial domains with some diffuse reflection at the boundary.

5.1.3 The Convergence Results for "Well-Prepared" Initial Data

Let us first recall from Chapter 2 that, at the present time, the mathematical theory of the Boltzmann equation is not really complete, insofar as there is no global existence and uniqueness result for general initial data with finite mass, energy and entropy.

We have therefore at our disposal either strong solutions with higher regularity which require smoothness and smallness assumptions on the initial data, or very weak solutions satisfying only a family of formally equivalent equations and thus called renormalized solutions. These renormalized solutions, built by DiPerna and Lions [44], exist globally in time without restriction on the size of the initial data but are not known to be unique, neither to satisfy the local conservations of momentum and energy.

In the framework of renormalized solutions, the strategy presented in the previous paragraph will not be completely effective, since the stability inequality for the modulated entropy relies crucially on the local conservation laws. We will thus restrict our attention to the particular case when the initial data is well-prepared, i.e. when $\rho_{in} = \theta_{in} = \nabla \cdot u_{in} = 0$. The approximate solution in (5.14) will therefore satisfy $T \equiv 1$ and we will not need anymore the local conservation of energy.

Theorem 5.1.8 *Let $(f_{\varepsilon,in})$ be a family of initial fluctuations around a global equilibrium M, i.e. satisfying*

$$\frac{1}{\varepsilon^2} H(f_{\varepsilon,in}|M) \leq C_{in}, \tag{5.15}$$

and converging entropically at order ε to $g_{in}(x,v) = u_{in}(x) \cdot v$

$$\frac{1}{\varepsilon^2} H(f_{\varepsilon,in}|\mathcal{M}_{1,\varepsilon u_{in},1}) \to 0 \ as \ \varepsilon \to 0, \tag{5.16}$$

for some given divergence-free vector field $u_{in} \in L^2(\Omega)$.

Let (f_ε) be a family of renormalized solutions to the scaled Boltzmann equation

$$\varepsilon \partial_t f_\varepsilon + v \cdot \nabla_x f_\varepsilon = \frac{1}{\varepsilon^q} Q(f_\varepsilon, f_\varepsilon) \text{ on } \mathbf{R}^+ \times \Omega \times \mathbf{R}^3,$$
$$f_\varepsilon(0, x, v) = f_{\varepsilon, in}(x, v) \text{ on } \Omega \times \mathbf{R}^3, \qquad (5.17)$$
$$f_\varepsilon(t, x, v) = f_\varepsilon(t, x, R_x v) \text{ on } \Sigma_-$$

for some $q > 1$, where Q is the collision operator defined by (2.7) associated with some collision kernel B satisfying Grad's cut off assumption (2.8).

Then the family of scaled bulk velocities (u_ε) defined by $u_\varepsilon = \frac{1}{\varepsilon} \int f_\varepsilon v dv$ is relatively weakly compact in $L^1_{loc}(dtdx)$, and any limit point u of (u_ε) is a dissipative solution to the incompressible Euler equations (5.1)(5.3).

If the limiting initial data u_{in} is smooth and such that (5.1) has a (unique) smooth solution u, the stability result above can be strengthened as follows, using the notion of entropic convergence defined in Chapter 4 :

Corollary 5.1.9 Let $(f_{\varepsilon, in})$ be a family of initial fluctuations around a global equilibrium M, converging entropically at order ε to $g_{in}(x, v) = u_{in}(x) \cdot v$ for some given $u_{in} \in H^s(\Omega)$ $(s > 5/2)$.

Let (f_ε) be a family of renormalized solutions to the scaled Boltzmann equation (5.17) for some $q > 1$, where Q is the collision operator defined by (2.7) associated with some collision kernel B satisfying Grad's cut off assumption (2.8).

Assume that the incompressible Euler equations (5.1) admit a strong solution $u \in L^\infty_{loc}([0, t^*], H^s(\Omega))$ such that

$$\int_0^{t^*} \|\nabla_x u + (\nabla_x u)^T\|_{L^2 \cap L^\infty(\Omega)}(t) dt < +\infty.$$

Then, for almost all $t \in [0, t^*]$, the family of fluctuations $(g_\varepsilon(t))$ defined by $f_\varepsilon = M(1 + \varepsilon g_\varepsilon)$ converges entropically to the infinitesimal Maxwellian $g(t)$ given by

$$g(t, x, v) = u(t, x) \cdot v.$$

Note that (5.16) is a very strong assumption on the family of initial data, meaning that *well-prepared initial data* has to be understood in the following sense.

- We first require that the initial distribution has a velocity profile close to local thermodynamic equilibrium

$$\rho_{in} + u_{in} \cdot v + \theta_{in} \frac{|v|^2 - 3}{2},$$

in order that there is no relaxation layer.
- We then ask the asymptotic initial thermodynamic fields to satisfy the incompressibility and Boussinesq constraints

$$\nabla \cdot u_{in} = 0, \quad \nabla(\rho_{in} + \theta_{in}) = 0,$$

which ensures that there is no acoustic wave. We further require that the initial temperature fluctuation (and thus mass fluctuation) is negligible

$$\rho_{in} = -\theta_{in} = 0 \,.$$

We therefore expect the temperature fluctuation to remain negligible.

- We finally need some spatial regularity on the limiting bulk velocity, more precisely we require some Lipschitz continuity.

We are thus able to consider very general initial data (satisfying only the physical estimate (5.15)), but in the vicinity of a small set of asymptotic distributions.

A natural question is then to know whether or not it is possible to get rid of these restrictions on the asymptotic distribution. In the proof that we will give in Sections 5.2 and 5.3, we will see that the first two assumptions come from the poor understanding of the Boltzmann equation, in particular from the fact that renormalized solutions to the Boltzmann equation are not known to satisfy the local conservation of energy (the heat flux is not even defined), whereas the last assumption concerning the regularity of the limiting distribution is inherent to the modulated entropy method.

Considering solutions to the Boltzmann equation satisfying rigorously the basic physical properties, we can actually expect to control the energy flux and extend the convergence result to take into account acoustic waves. In order to also deal with the relaxation layer, we further need to understand the dissipation mechanism, which can be done by slight modifications of the method.

On the contrary, relaxing the regularity assumption requires new ideas : the stability in energy and entropy methods is indeed controlled by the Lipschitz norm of the limiting field. In 3D, the incompressible Euler equations are not known to have weak solutions, so that we do not expect to extend our convergence result for distributions with lower regularity. In return, in 2D, the mathematical theory of the incompressible Euler equations is much better understood and singular solutions such as vortex patches are known to exist globally in time : it should be then relevant to study the hydrodynamic limit of the Boltzmann equation in this setting. By analogy with the compressible Euler equations, we would expect the spatial discontinuities to dissipate entropy, or in other words to create layers where the distribution is far from local thermodynamic equilibrium. The difficulty should be to split the space-time domain according to these layers.

5.1.4 The Convergence Result for General Initial Data

The second result we will state here answers the previous question in the case of smooth limiting fields. Considering a stronger notion of solution for the Boltzmann equation (5.17), for instance using the classical solutions built by

Guo [62], we can prove the convergence to the incompressible Euler equations for general initial data.

We indeed recall that nonlinear energy methods allow to build global smooth solutions to the Boltzmann equation for small data (see [62]) :

Proposition 5.1.10 *Consider the collision cross-section B of hard spheres. Given any initial data f_{in} satisfying*

$$\left\| (1 + |v|)^{1/2} D_x^s \left(\frac{f_{in} - M}{\sqrt{M}} \right) \right\|_{L^2(\mathbf{R}^3 \times \mathbf{R}^3)} \leq \delta$$

for $s \geq 4$ and δ sufficiently small, there exists a unique classical solution f to the Boltzmann equation (5.17) with initial data f_{in} (such that the previous norm remains bounded for all time). In particular it satisfies the local conservation laws as well as the local entropy inequality.

Note that, for our asymptotic study, we do not need so much regularity : we will only require that the solutions of the Boltzmann equation to be considered satisfy the non uniform nonlinear estimate

$$\frac{1}{\varepsilon^2} \int M \left(\frac{f_\varepsilon - M}{M} \right)^2 dv \leq \frac{C}{\varepsilon^2} \text{ a.e. on } \mathbf{R}^+ \times \Omega$$

The previous proposition just ensures that such solutions exist.

Theorem 5.1.11 *Let $(f_{\varepsilon,in})$ be a family of nonnegative functions on $\Omega \times \mathbf{R}^3$ satisfying the scaling condition (5.15)*

$$\frac{1}{\varepsilon^2} H(f_{\varepsilon,in}|M) \leq C_{in}.$$

Without loss of generality, we further assume that the fluctuations $(g_{\varepsilon,in})$ defined by $f_{\varepsilon,in} = M(1 + \varepsilon g_{\varepsilon,in})$ converge entropically (of order ε) to some g_{in}:

$$\frac{1}{\varepsilon^2} H(f_{\varepsilon,in}|M) \rightarrow \frac{1}{2} \iint M g_{in}^2 dv dx. \tag{5.18}$$

Let (f_ε) be some family of solutions to the scaled Boltzmann equation (5.17) with $q > 1$, satisfying further

$$\int M \left(\frac{f_\varepsilon - M}{M} \right)^2 dv \leq C \text{ a.e. on } \mathbf{R}^+ \times \Omega. \tag{5.19}$$

Then, up to extraction of a subsequence, the family of fluctuations (g_ε) defined by $f_\varepsilon = M(1 + \varepsilon g_\varepsilon)$ converges weakly to $u \cdot v + \frac{1}{2}\theta \left(|v|^2 - 5 \right)$, where (u, θ) is the Lipschitz solution to the incompressible Euler equations

$$\begin{aligned} \partial_t u + u \cdot \nabla_x u + \nabla_x p = 0, \quad \nabla_x \cdot u = 0 \text{ on } \mathbf{R}^+ \times \Omega, \\ \partial_t \theta + u \cdot \nabla_x \theta = 0 \text{ on } \mathbf{R}^+ \times \Omega, \\ u(0, x) = P u_{in}(x), \quad \theta(0, x) = \frac{1}{5}(3\theta_{in} - 2\rho_{in}) \text{ on } \Omega, \end{aligned} \tag{5.20}$$

as long as the latter does exist.

Furthermore the difference $g_\varepsilon - g$ behaves asymptotically in L^1 as

$$g_{osc}(\frac{t}{\varepsilon}, x, v) = (\rho_{osc}, u_{osc}, \theta_{osc})(\frac{t}{\varepsilon}, x) \cdot \left(1, v, \frac{1}{2}(|v|^2 - 3)\right)$$

where $(\rho_{osc}, u_{osc}, \theta_{osc})$ is the solution of the acoustic system (5.45) stated in Section 5.4.1.

Note that the purely kinetic part does not appear in that convergence statement since its contribution to the L^1 norm is negligible. The entropic convergence we will establish is actually stronger.

5.2 The Relative Entropy Method

5.2.1 Description of the Strategy

Our goal here is to establish the convergence of appropriately scaled families of solutions to the Boltzmann equation towards solutions of the incompressible Euler equations.

The first result in the framework of renormalized solutions is due to Golse [16], and is based on the *relative entropy method*, which can be sketched as follows :

- In terms of g_ε, the Boltzmann equation (5.17) become

$$\varepsilon^q \partial_t g_\varepsilon + \varepsilon^{q-1} v \cdot \nabla_x g_\varepsilon + \frac{1}{\varepsilon} \mathcal{L}_M g_\varepsilon = \frac{1}{M} Q(M g_\varepsilon, M g_\varepsilon), \qquad (5.21)$$

  where $-\mathcal{L}_M$ denotes as previously the linearization of Boltzmann's collision integral at the Maxwellian state M. Therefore, multiplying (5.21) by ε and letting $\varepsilon \to 0$ suggests that g_ε converges in the sense of distributions to some *infinitesimal Maxwellian* g (see Proposition 3.2.2 in Chapter 3) :

$$g(t, x, v) = \rho(t, x) + u(t, x) \cdot v + \frac{1}{2}\theta(t, x)(|v|^2 - 3).$$

- Passing to the limit in the local conservations of mass and momentum leads then to

$$\nabla_x \cdot u = 0 \text{ and } \nabla_x(\rho + \theta) = 0,$$

  known as *incompressibility and Boussinesq relations*.
- The core of the proof is therefore to establish a stability inequality on the modulated entropy of the same type as (5.11) :

$$\frac{1}{\varepsilon^2} H\left(f_\varepsilon | \mathcal{M}_{(\exp(\varepsilon\tilde\rho),\varepsilon\tilde u,\exp(\varepsilon\tilde\theta))}\right)(t) + \frac{1}{\varepsilon^{q+3}} \int_0^t \int D(f_\varepsilon) ds dx$$

$$\leq \frac{1}{\varepsilon^2} H\left(f_{\varepsilon,in} | \mathcal{M}_{(\exp(\varepsilon\tilde\rho_{in}),\varepsilon\tilde u_{in},\exp(\varepsilon\tilde\theta_{in}))}\right) + \frac{1}{\varepsilon^2} \int_0^t \int \partial_t \exp(\varepsilon\tilde\rho) dx ds$$

$$- \frac{1}{\varepsilon} \int_0^t \iint f_\varepsilon \left(1, e^{-\varepsilon\tilde\theta}(v - \varepsilon\tilde u), \frac{1}{2}\left(\frac{|v - \varepsilon\tilde u|^2}{e^{\varepsilon\tilde\theta}} - 3\right)\right)$$
$$\cdot \mathbf{A}_\varepsilon(\tilde\rho, \tilde u, \tilde\theta) dv dx ds$$

$$- \frac{1}{\varepsilon^2} \int_0^t \iint f_\varepsilon \left(\nabla_x \tilde u : \Phi_\varepsilon + e^{\frac{1}{2}\varepsilon\tilde\theta}\nabla_x\tilde\theta \cdot \Psi_\varepsilon\right) dx dv ds$$

denoting by $\mathbf{A}_\varepsilon(\rho, u, \theta)$ the acceleration operator (which differs from the incompressible Euler equations with temperature (5.20) by some penalization describing the acoustic waves), and by Φ_ε and Ψ_ε the kinetic momentum and energy fluxes - which are scaled translated variants of

$$\Phi = \left(v^{\otimes 2} - \frac{1}{3}|v|^2 Id\right),$$
$$\Psi = \frac{1}{2}v\left(|v|^2 - 5\right).$$

We conclude then the proof of Theorem 5.1.8 by remarking that the relative entropy $\varepsilon^{-2} H(f_\varepsilon | \mathcal{M}_{(\exp(\varepsilon\tilde\rho),\varepsilon\tilde u,\exp(\varepsilon\tilde\theta))})$ controls asymptotically the L^2 norm of $(\rho - \tilde\rho)$, $(u - \tilde u)$ and $(\theta - \tilde\theta)$.

- In the case when the incompressible Euler equations admit a smooth solution u, we have further

$$\frac{1}{\varepsilon^2} H\left(f_\varepsilon | \mathcal{M}_{(\exp(\varepsilon\rho),\varepsilon u,\exp(\varepsilon\theta))}\right) \to 0$$

just by choosing $\tilde\rho = \rho$, $\tilde u = u$ and $\tilde\theta = \theta$ in the previous stability inequality, and thus the expected entropic convergence.

In [16], the convergence of renormalized solutions of the scaled Boltzmann equation to solutions of the incompressible Euler equations is established assuming

(i) the initial data to be well-prepared

$$\rho_{in} = \theta_{in} = \nabla \cdot u_{in} = 0,$$

and thus choosing only test fields such that

$$\tilde\rho = \tilde\theta = \nabla \cdot \tilde u = 0;$$

(ii) some nonlinear estimate, namely

$$(1 + |v|^2)\frac{g_\varepsilon^2}{1 + \frac{\varepsilon}{2}g_\varepsilon} \text{ relatively weakly compact in } L^1_{loc}(dtdx, w - L^1(Mdv))$$

which provides both a control on large velocities v, and some equiintegrability with respect to space variables x; and

(iii) the local conservation of momentum which is not guaranteed for renormalized solutions of the Boltzmann equation — see Chapter 2 for a discussion of this particular point.

Assumption (iii) was removed by Lions and Masmoudi in [75]; their argument uses the local conservation of momentum with matrix-valued defect measure satisfied by renormalized solutions of the Boltzmann equation (see Theorem 2.3.4 in Chapter 2). That this defect measure vanishes in the incompressible Euler limit will follow from the strong convergence result to be proved.

Another, more serious difficulty is to circumvent the need for assumption (ii). Indeed we cannot hope to establish such a statement since, in incompressible inviscid regime, even at formal level, we have no control on the transport $(\varepsilon \partial_t + v \cdot \nabla_x) g_\varepsilon$, and therefore no spatial regularity (as for the target equations). The idea is to introduce a suitable decomposition of the momentum flux, and to estimate each term in that decomposition either by the modulated entropy, or by the entropy dissipation. In other words, the argument is based on loop estimates instead of a priori estimates, and the conclusion follows from Gronwall's inequality. This strategy based on Gronwall's inequality has been first used in the framework of the BGK equation [93], and then adapted to the original Boltzmann equation [94] using refined dissipation estimates from [54] and [55] discussed in Chapter 3.

Assumption (i) has been removed by the author in [95] under an additional integrability condition on the solutions to the scaled Boltzmann equation. It will be discussed in the last two sections of this chapter. In Section 5.4, we will indeed describe the (fast oscillating) acoustic waves arising when the thermodynamic fields are not well-prepared, i.e. when the incompressibility and Boussinesq constraints

$$\nabla \cdot u = 0, \quad \nabla(\rho + \theta) = 0$$

are not satisfied initially. In Section 5.5, we will then deal with the relaxation layer arising when the initial velocity profile is not close to local thermodynamic equilibrium, i.e. when

$$\frac{1}{\varepsilon^2} H\big(f_{\varepsilon,in} | \mathcal{M}_{(\exp(\varepsilon \rho_{in}), \varepsilon u_{in}, \exp(\varepsilon \theta_{in}))}\big)$$

does not converge to 0. In both cases, we will build refined approximate solutions such that the modulated entropy converges strongly to 0. Let us however point out that the convergence proof, in particular the estimate on the flux terms, requires some additional integrability condition on the solutions to the scaled Boltzmann equation.

The suitable functional framework to study the Euler asymptotics is therefore slightly different from that defined in the previous chapters. Instead of defining the fluctuation relative to the global equilibrium M, we will consider the fluctuation relative to the local equilibrium

$$\mathcal{M}_\varepsilon \stackrel{\text{def}}{=} \mathcal{M}_{(\exp(\varepsilon\tilde{\rho}),\varepsilon\tilde{u},\exp(\varepsilon\tilde{\theta}))}$$

occuring in the modulated entropy, and use the properties of the linearized collision operator $\mathcal{L}_\varepsilon$ defined by

$$\mathcal{L}_\varepsilon g = -\frac{2}{\mathcal{M}_\varepsilon} Q\left(\mathcal{M}_\varepsilon, \mathcal{M}_\varepsilon g\right), \tag{5.22}$$

which can be easily deduced from the properties of $\mathcal{L}_M$ by translation.

Let us therefore introduce the suitable corresponding notations and basic estimates obtained for the scaled Boltzmann equation (5.17). Two important quantities are the renormalized fluctuation $\tilde{g}_\varepsilon$ and the renormalized collision integral $\tilde{q}_\varepsilon$ defined by

$$\begin{aligned}
\tilde{g}_\varepsilon &= \frac{2}{\varepsilon}\left(\sqrt{\frac{f_\varepsilon}{\mathcal{M}_\varepsilon}} - 1\right), \\
\tilde{q}_\varepsilon &= \frac{1}{\varepsilon^{(q+3)/2}}\frac{1}{\mathcal{M}_\varepsilon}Q(\sqrt{\mathcal{M}_\varepsilon f_\varepsilon}, \sqrt{\mathcal{M}_\varepsilon f_\varepsilon})
\end{aligned} \tag{5.23}$$

for which the modulated entropy and the entropy dissipation give the following L^2 bounds

$$\begin{aligned}
\|\tilde{g}_\varepsilon\|^2_{L^\infty(\mathbf{R}^+, L^2(\mathcal{M}_\varepsilon dvdx))} &\le \frac{2}{\varepsilon^2}H(f_\varepsilon|\mathcal{M}_\varepsilon), \\
\|\tilde{q}_\varepsilon\|^2_{L^2(dtdx\nu_\varepsilon^{-1}\mathcal{M}_\varepsilon dv)} &\le \frac{1}{\varepsilon^{q+3}}\iint D(f_\varepsilon)(t,x)dtdx \le C_{in}
\end{aligned} \tag{5.24}$$

By some variant of Lemma 3.2.4 we further have the following relaxation bound

$$\|\tilde{g}_\varepsilon - \Pi_\varepsilon \tilde{g}_\varepsilon\|_{L^2(\mathcal{M}_\varepsilon dv)} = O(\varepsilon^{(q+1)/2})_{L^2_{t,x}} + O(\varepsilon)\|\tilde{g}_\varepsilon\|^2_{L^2(\mathcal{M}_\varepsilon dv)}, \tag{5.25}$$

where Π_ε denotes the orthogonal projection on the kernel of $\mathcal{L}_\varepsilon$.

5.2.2 The Stability Inequality in the Framework of Renormalized Solutions

In order to establish the stability inequality leading to the entropic convergence stated in Theorem 5.1.8, the starting point is the derivation of the modulated entropy :

Proposition 5.2.1 *For any divergence free vector-field $\tilde{u} \in C_c^\infty(\mathbf{R}^+ \times \Omega)$ such that $n \cdot \tilde{u}_{|\partial\Omega} = 0$, denote*

$$\mathcal{M}_\varepsilon \stackrel{\text{def}}{=} \mathcal{M}_{(1,\varepsilon\tilde{u},1)}.$$

Then, under the same assumptions as in Theorem 5.1.8, one has

$$\frac{1}{\varepsilon^2}H(f_\varepsilon|\mathcal{M}_\varepsilon)(t) + \frac{1}{\varepsilon^2}\int \mathrm{tr}(m_\varepsilon)(t) + \frac{1}{\varepsilon^{q+3}}\int_0^t\int D(f_\varepsilon)(s,x)dxds$$

$$\leq \frac{1}{\varepsilon^2}H(f_{\varepsilon,in}|\mathcal{M}_{\varepsilon,in}) + \frac{1}{\varepsilon}\int_0^t\int A(\tilde{u})\cdot\int(\varepsilon\tilde{u}-v)f_\varepsilon(s,x,v)dvdxds$$

$$-\frac{1}{2\varepsilon^2}\int_0^t\int(\nabla_x\tilde{u}+(\nabla_x\tilde{u})^T):\left(m_\varepsilon(s)+\int(v-\varepsilon\tilde{u})^{\otimes 2}f_\varepsilon(s,x,v)dvdx\right)ds$$

(5.26)

with

$$A(\tilde{u}) = \partial_t\tilde{u} + (\tilde{u}\cdot\nabla_x)\tilde{u}. \tag{5.27}$$

Proof. Let us first recall the entropy inequality with defect measure (2.36) satisfied by renormalized solutions of the scaled Boltzmann equation (5.17) with specular reflection at the boundary:

$$H(f_\varepsilon(t)|M) + \int_{\mathbf{R}^3}\mathrm{tr}(m_\varepsilon)(t) + \frac{1}{\varepsilon^{q+1}}\int_0^t\int D(f_\varepsilon)(s,x)dsdx \leq H(f_{\varepsilon,in}|M)$$

(5.28)

where $m_\varepsilon \in L^\infty(\mathbf{R}^+,\mathcal{M}(\Omega,M_3(\mathbf{R})))$ is the momentum defect measure.

By definition of the modulated entropy (5.14), we then have

$$H(f_\varepsilon|\mathcal{M}_\varepsilon)(t) + \int_{\mathbf{R}^3}\mathrm{tr}(m_\varepsilon)(t) + \frac{1}{\varepsilon^{q+1}}\int_0^t\int D(f_\varepsilon)dsdx$$

$$\leq H(f_{\varepsilon,in}|\mathcal{M}_{\varepsilon,in}) + \int_0^t\frac{d}{dt}\iint\frac{1}{2}(\varepsilon^2\tilde{u}^2 - 2\varepsilon v\cdot\tilde{u})f_\varepsilon(s,x,v)dvdxds$$

From the continuity equation

$$\partial_t\int f_\varepsilon dv + \nabla_x\cdot\frac{1}{\varepsilon}\int vf_\varepsilon dv = 0, \tag{5.29}$$

and the conservation of momentum with defect measure

$$\partial_t\int vf_\varepsilon dv + \nabla_x\cdot\frac{1}{\varepsilon}\int v\otimes vf_\varepsilon dv + \frac{1}{\varepsilon}\nabla_x\cdot m_\varepsilon = 0, \tag{5.30}$$

we deduce that

$$H(f_\varepsilon|\mathcal{M}_\varepsilon)(t) + \int_{\mathbf{R}^3}\mathrm{tr}(m_\varepsilon)(t) + \frac{1}{\varepsilon^{q+1}}\int_0^t\int D(f_\varepsilon)dsdx$$

$$\leq H(f_{\varepsilon,in}|\mathcal{M}_{\varepsilon,in}) + \int_0^t\iint\varepsilon\partial_t\tilde{u}\cdot(\varepsilon\tilde{u}-v)f_\varepsilon(s,x,v)dvdxds$$

$$-\frac{1}{2}\int_0^t\int\varepsilon\tilde{u}^2(s,x)\nabla_x\cdot\left(\int vf_\varepsilon(s,x,v)dv\right)dxds$$

$$+\int_0^t\int\tilde{u}\cdot\left(\nabla_x\cdot\left(\int v\otimes vf_\varepsilon(s,x,v)dvdx + m_\varepsilon(s,x)\right)\right)ds$$

Integrating by parts (using the zero mass flux condition on $\tilde{u}$ and the specular reflection for f_ε), we get

$$
H(f_\varepsilon|\mathcal{M}_\varepsilon)(t) + \int_{\mathbf{R}^3} \mathrm{tr}(m_\varepsilon)(t) + \frac{1}{\varepsilon^{q+1}} \int_0^t \int D(f_\varepsilon) ds dx
$$
$$
\leq H(f_{\varepsilon,in}|\mathcal{M}_{\varepsilon,in}) + \int_0^t \iint \varepsilon \partial_t \tilde{u} \cdot (\varepsilon \tilde{u} - v) f_\varepsilon(s,x,v) dv dx ds
$$
$$
+ \int_0^t \int \varepsilon \nabla_x \tilde{u} : \tilde{u} \otimes \left(\int v f_\varepsilon(s,x,v) dv \right) dx ds
$$
$$
- \int_0^t \int \nabla_x \tilde{u} : \left(\int v \otimes v f_\varepsilon(s,x,v) dv dx + m_\varepsilon(s,x) \right) ds
$$

This clearly implies

$$
H(f_\varepsilon|\mathcal{M}_\varepsilon)(t) + \int_{\mathbf{R}^3} \mathrm{tr}(m_\varepsilon)(t) + \frac{1}{\varepsilon^{q+1}} \int_0^t \int D(f_\varepsilon) ds dx
$$
$$
\leq H(f_{\varepsilon,in}|\mathcal{M}_{\varepsilon,in}) + \int_0^t \iint \varepsilon \partial_t \tilde{u} \cdot (\varepsilon \tilde{u} - v) f_\varepsilon(s,x,v) dv dx ds
$$
$$
- \int_0^t \int \nabla_x \tilde{u} : \left(\int (v - \varepsilon \tilde{u}) \otimes v f_\varepsilon(s,x,v) dv dx + m_\varepsilon(s,x) \right) ds
$$

or equivalently

$$
H(f_\varepsilon|\mathcal{M}_\varepsilon)(t) + \int_{\mathbf{R}^3} \mathrm{tr}(m_\varepsilon)(t) + \frac{1}{\varepsilon^{q+1}} \int_0^t \int D(f_\varepsilon) ds dx
$$
$$
\leq H(f_{\varepsilon,in}|\mathcal{M}_{\varepsilon,in}) + \int_0^t \iint \varepsilon \partial_t \tilde{u} \cdot (\varepsilon \tilde{u} - v) f_\varepsilon(s,x,v) dv dx ds
$$
$$
- \int_0^t \int \nabla_x \tilde{u} : \left(\int (v - \varepsilon \tilde{u})^{\otimes 2} f_\varepsilon(s,x,v) dv dx + m_\varepsilon(s,x) \right) dx ds
$$
$$
- \int_0^t \int \varepsilon \nabla_x \tilde{u} : \left(\int (v - \varepsilon \tilde{u}) \otimes \tilde{u} f_\varepsilon(s,x,v) dv \right) ds
$$

Introducing the acceleration operator $A(\tilde{u})$ leads therefore to the expected inequality. $\qquad\square$

5.2.3 The Stability Inequality Under the Additional Integrability Assumption

Under the assumption (5.19) of Theorem 5.1.11, the previous modulated entropy inequality can be extended as follows :

Proposition 5.2.2 *Denote by $\mathcal{M}_\varepsilon$ the fluctuation of Maxwellian defined by*

$$
\mathcal{M}_\varepsilon \overset{def}{=} \mathcal{M}_{(\exp(\varepsilon\tilde\rho), \varepsilon\tilde{u}, \exp(\varepsilon\tilde\theta))}
$$

for any $(\tilde\rho, \tilde{u}, \tilde\theta) \in C_c^\infty(\mathbf{R}^+ \times \Omega)$.

Then, any solution to the scaled Boltzmann equation (5.17) such that (5.19) holds, satisfies the following modulated entropy inequality

$$H\left(f_\varepsilon | \mathcal{M}_\varepsilon\right)(t) + \frac{1}{\varepsilon^{q+1}} \int_0^t \int D(f_\varepsilon) ds dx$$

$$\leq H(f_{\varepsilon,in} | \mathcal{M}_{\varepsilon,in}) + \int_0^t \int \partial_t \exp(\varepsilon\tilde{\rho}) dx ds$$

$$- \varepsilon \int_0^t \iint f_\varepsilon \left(1, e^{-\varepsilon\tilde{\theta}}(v - \varepsilon\tilde{u}), \frac{1}{2}\left(e^{-\varepsilon\tilde{\theta}}|v - \varepsilon\tilde{u}|^2 - 3\right)\right) \cdot \mathbf{A}_\varepsilon(\tilde{\rho}, \tilde{u}, \tilde{\theta}) dv dx ds$$

$$- \int_0^t \iint f_\varepsilon \nabla_x \tilde{u} : \Phi_\varepsilon dx dv + \iint f_\varepsilon e^{\frac{1}{2}\varepsilon\tilde{\theta}} \nabla_x \tilde{\theta} \cdot \Psi_\varepsilon dx dv ds$$

$$(5.31)$$

for some acceleration operator $\mathbf{A}_\varepsilon(\tilde{\rho}, \tilde{u}, \tilde{\theta})$ to be defined by (5.35).

Proof. Start from the entropy inequality satisfied by the solution of the scaled Boltzmann equation with specular reflection at the boundary:

$$H(f_\varepsilon(t)|M) + \frac{1}{\varepsilon^{q+1}} \int_0^t \int D(f_\varepsilon)(s, x) ds dx \leq H(f_\varepsilon^{in}|M) \qquad (5.32)$$

By definition of the modulated entropy and of the approximate solution $\mathcal{M}_\varepsilon$, we then have

$$H\left(f_\varepsilon | \mathcal{M}_\varepsilon\right)(t) + \frac{1}{\varepsilon^{q+1}} \int_0^t \int D(f_\varepsilon) ds dx$$

$$\leq H(f_{\varepsilon,in}|\mathcal{M}_{\varepsilon,in}) + \int_0^t \int \partial_t \left(\int \mathcal{M}_\varepsilon dv\right) dx ds \qquad (5.33)$$

$$- \int_0^t \frac{d}{dt} \iint \left(\varepsilon\left(\tilde{\rho} - \frac{3}{2}\tilde{\theta}\right) - \frac{1}{2}e^{-\varepsilon\tilde{\theta}}|v - \varepsilon\tilde{u}|^2 + \frac{1}{2}|v|^2\right) f_\varepsilon dv dx ds$$

with

$$\int \mathcal{M}_\varepsilon dv = \exp(\varepsilon\tilde{\rho}).$$

Now, let us point out that, under assumption (5.19), f_ε satisfies the local conservation laws. Then, using the continuity equation

$$\partial_t \int f_\varepsilon dv + \nabla_x \cdot \frac{1}{\varepsilon} \int v f_\varepsilon dv = 0,$$

the conservation of momentum

$$\partial_t \int v f_\varepsilon dv + \nabla_x \cdot \frac{1}{\varepsilon} \int v \otimes v f_\varepsilon dv = 0,$$

and the conservation of energy

$$\partial_t \int \frac{1}{2}|v|^2 f_\varepsilon dv + \nabla_x \cdot \frac{1}{\varepsilon} \int \frac{1}{2}|v|^2 v f_\varepsilon dv = 0,$$

as well as the boundary condition on Σ_-

$$f_\varepsilon(t, x, v) = f_\varepsilon(t, x, v - 2(v \cdot n(x))n(x)),$$

and integrating by parts, we obtain

$$\frac{1}{\varepsilon}\frac{d}{dt}\iint\left(\varepsilon\left(\tilde\rho - \frac{3}{2}\tilde\theta\right) - \frac{1}{2}e^{-\varepsilon\tilde\theta}|v - \varepsilon\tilde u|^2 + \frac{1}{2}|v|^2\right)f_\varepsilon \, dvdx$$

$$= \iint f_\varepsilon\left(\partial_t\left(\tilde\rho - \frac{3}{2}\tilde\theta\right) + (\tilde u \cdot \nabla_x)\left(\tilde\rho - \frac{3}{2}\tilde\theta\right) + \frac{1}{\varepsilon}(v - \varepsilon\tilde u) \cdot \nabla_x\left(\tilde\rho - \frac{3}{2}\tilde\theta\right)\right)dvdx$$

$$+ \iint f_\varepsilon e^{-\varepsilon\tilde\theta}(v - \varepsilon\tilde u) \cdot \left(\partial_t\tilde u + (\tilde u \cdot \nabla_x)\tilde u + \frac{1}{\varepsilon}(v - \varepsilon\tilde u) \cdot \nabla_x\tilde u\right)dvdx$$

$$+ \frac{1}{2}\iint f_\varepsilon e^{-\varepsilon\tilde\theta}|v - \varepsilon\tilde u|^2\left(\partial_t\tilde\theta + (\tilde u \cdot \nabla_x)\tilde\theta + \frac{1}{\varepsilon}(v - \varepsilon\tilde u) \cdot \nabla_x\tilde\theta\right)dvdx$$

provided that $\tilde u \cdot n = 0$ on $\partial\Omega$.

Let us then introduce the kinetic momentum and energy fluxes

$$\Phi_\varepsilon = e^{-\varepsilon\tilde\theta}\left((v - \varepsilon\tilde u)^{\otimes 2} - \frac{1}{3}|v - \varepsilon\tilde u|^2 Id\right),$$

$$\Psi_\varepsilon = \frac{1}{2}e^{-\frac{3}{2}\varepsilon\tilde\theta}(v - \varepsilon\tilde u)\left(|v - \varepsilon\tilde u|^2 - 5e^{\varepsilon\tilde\theta}\right) \tag{5.34}$$

and recall that Φ_ε and Ψ_ε belong to the orthogonal complement of the kernel $\mathrm{Ker}\mathcal{L}_\varepsilon$ where $\mathcal{L}_\varepsilon$ is the linearized collision operator at $\mathcal{M}_\varepsilon$. We have

$$e^{-\varepsilon\tilde\theta}\nabla_x\tilde u : (v - \varepsilon\tilde u)^{\otimes 2} = \nabla_x\tilde u : \Phi_\varepsilon + \frac{1}{3}e^{-\varepsilon\tilde\theta}\nabla_x \cdot \tilde u|v - \varepsilon\tilde u|^2$$

$$\frac{1}{2}e^{-\varepsilon\tilde\theta}\nabla_x\tilde\theta \cdot (v - \varepsilon\tilde u)|v - \varepsilon\tilde u|^2 = e^{\frac{1}{2}\varepsilon\tilde\theta}\nabla_x\tilde\theta \cdot \Psi_\varepsilon + \frac{5}{2}\nabla_x\tilde\theta \cdot (v - \varepsilon\tilde u)$$

so that

$$\frac{1}{\varepsilon}\frac{d}{dt}\iint\left(\varepsilon\left(\tilde\rho - \frac{3}{2}\tilde\theta\right) - \frac{1}{2}e^{-\varepsilon\tilde\theta}|v - \varepsilon\tilde u|^2 + \frac{1}{2}|v|^2\right)f_\varepsilon \, dvdx$$

$$= \iint f_\varepsilon\left(\partial_t\left(\tilde\rho - \frac{3}{2}\tilde\theta\right) + (\tilde u \cdot \nabla_x)\left(\tilde\rho - \frac{3}{2}\tilde\theta\right)\right)dvdx$$

$$+ \iint f_\varepsilon e^{-\varepsilon\tilde\theta}(v - \varepsilon\tilde u) \cdot \left(\partial_t\tilde u + (\tilde u \cdot \nabla_x)\tilde u + \frac{1}{\varepsilon}e^{\varepsilon\tilde\theta}\nabla_x\left(\tilde\rho - \frac{3}{2}\tilde\theta\right) + \frac{5}{2\varepsilon}e^{\varepsilon\tilde\theta}\nabla_x\tilde\theta\right)dvdx$$

$$+ \frac{1}{2}\iint f_\varepsilon e^{-\varepsilon\tilde\theta}|v - \varepsilon\tilde u|^2\left(\partial_t\tilde\theta + (\tilde u \cdot \nabla_x)\tilde\theta + \frac{2}{3}\nabla_x \cdot \tilde u\right)dvdx$$

$$+ \frac{1}{\varepsilon}\iint f_\varepsilon \nabla_x\tilde u : \Phi_\varepsilon \, dxdv + \frac{1}{\varepsilon}\iint f_\varepsilon e^{\frac{1}{2}\varepsilon\tilde\theta}\nabla_x\tilde\theta \cdot \Psi_\varepsilon \, dxdv.$$

It is then natural to define the acceleration operator

$$A_\varepsilon(\rho, u, \theta) = \begin{pmatrix} \partial_t\rho + (u \cdot \nabla_x)\rho + \dfrac{1}{\varepsilon}\nabla_x \cdot u \\[2mm] \partial_t u + (u \cdot \nabla_x)u + \left(\dfrac{e^{\varepsilon\theta} - 1}{\varepsilon}\right)\nabla_x(\rho + \theta) + \dfrac{1}{\varepsilon}\nabla_x(\rho + \theta) \\[2mm] \partial_t\theta + (u \cdot \nabla_x)\theta + \dfrac{2}{3\varepsilon}\nabla_x \cdot u \end{pmatrix} \tag{5.35}$$

so that the inequality can be recasted in suitable form

$$\frac{1}{\varepsilon}\frac{d}{dt}\iint\left(\varepsilon\left(\tilde{\rho}-\frac{3}{2}\tilde{\theta}\right)-\frac{1}{2}e^{-\varepsilon\tilde{\theta}}|v-\varepsilon\tilde{u}|^2+\frac{1}{2}|v|^2\right)f_\varepsilon dvdx$$
$$=\iint f_\varepsilon\left(1,e^{-\varepsilon\tilde{\theta}}(v-\varepsilon\tilde{u}),\frac{1}{2}\left(e^{-\varepsilon\tilde{\theta}}|v-\varepsilon\tilde{u}|^2-3\right)\right)\cdot A_\varepsilon(\tilde{\rho},\tilde{u},\tilde{\theta})dvdx$$
$$+\frac{1}{\varepsilon}\iint f_\varepsilon\nabla_x\tilde{u}:\Phi_\varepsilon dxdv+\frac{1}{\varepsilon}\iint f_\varepsilon e^{\frac{1}{2}\varepsilon\tilde{\theta}}\nabla_x\tilde{\theta}\cdot\Psi_\varepsilon dxdv$$

Plugging this last inequality in (5.33) leads to the announced result.

Note that the acceleration operator defined by (5.35) differs from the previous one (5.27) (defined for well-prepared initial data) by some penalization forcing the weak limit to satisfy the constraints

$$\nabla_x\cdot u=0,\quad\nabla_x(\rho+\theta)=0.$$

$\square$

5.3 The Case of "Well-Prepared" Initial Data

We first focus on the case of "well-prepared" initial data. In that case, we have only to consider test Maxwellians

$$\mathcal{M}_\varepsilon=\mathcal{M}_{(\exp(\varepsilon\tilde{\rho},\varepsilon\tilde{u},\exp(\varepsilon\tilde{\theta}))}\text{ with }\tilde{\rho}=\tilde{\theta}=\nabla\cdot\tilde{u}=0.$$

We have then, considering any renormalized solution to (5.17)

$$\frac{1}{\varepsilon^2}H(f_\varepsilon|\mathcal{M}_\varepsilon)(t)+\frac{1}{\varepsilon^2}\int\text{tr}(m_\varepsilon)(t)+\frac{1}{\varepsilon^{q+3}}\int_0^t\iint D(f_\varepsilon)(s,x)dxds$$
$$\leq\frac{1}{\varepsilon^2}H(f_{\varepsilon,in}|\mathcal{M}_{\varepsilon,in})+\frac{1}{\varepsilon}\int_0^t\int A(\tilde{u})\cdot\int(\varepsilon\tilde{u}-v)f_\varepsilon(s,x,v)dvdxds$$
$$-\frac{1}{2\varepsilon^2}\int_0^t\int(\nabla_x\tilde{u}+(\nabla_x\tilde{u})^T):\left(m_\varepsilon(s)+\int\Phi_\varepsilon f_\varepsilon(s,x,v)dvdx\right)ds$$

with

$$\Phi_\varepsilon(v)=(v-\varepsilon\tilde{u})^{\otimes2}-\frac{1}{3}|v-\varepsilon\tilde{u}|^2\,\text{Id}\,.$$

5.3.1 Control of the Flux Term

We expect to deduce from that inequality the convenient stability inequality, provided that we are able to deal with the flux term, namely with

$$-\frac{1}{2\varepsilon^2}\int_0^t\iint(\nabla_x\tilde{u}+(\nabla_x\tilde{u})^T):\Phi_\varepsilon f_\varepsilon(s,x,v)dvdxds$$

Passing to the limit directly in the flux term requires a priori estimates that control the effect of large velocities v and entail equiintegrability with respect to x variables. Such a method fails therefore with this type of scaling since no uniform spatial regularity is known to hold a priori.

The key idea is then to estimate the flux in terms of the modulated entropy and of the entropy dissipation, and then to conclude by Gronwall's lemma. We have

Proposition 5.3.1 *Under the same assumptions as in Theorem 5.1.8, for all solenoidal vector field $\tilde{u} \in C_c^\infty([0, t^*] \times \bar{\Omega})$,*

$$-\frac{1}{2\varepsilon^2} \int_0^t \iint (\nabla_x \tilde{u} + (\nabla_x \tilde{u})^T) : \Phi_\varepsilon(f_\varepsilon - M_\varepsilon)(s, x, v) dv dx ds$$

$$\leq \frac{C}{\varepsilon^2} \int_0^t \|(\nabla_x \tilde{u} + (\nabla_x \tilde{u})^T)\|_{L^2 \cap L^\infty(\Omega)} H(f_\varepsilon | M_\varepsilon)(s) ds + o(1)$$

(5.36)

Proof. The main idea behind this result is that the local thermodynamic equilibrium M_{f_ε} is expected to give a good approximation of the distribution f_ε in the fast relaxation limit, at least if the moments remain bounded. Now, for Maxwellian distributions, i.e. if $f_\varepsilon = M_{f_\varepsilon}$, the flux term can be computed explicitly in terms of the moments and estimated by the modulated entropy $H(M_{f_\varepsilon} | M_\varepsilon)$ which is more or less equivalent to the L^2 norm of the moments of $M_{f_\varepsilon} - M_\varepsilon$:

$$H(M_{f_\varepsilon} | M_\varepsilon) = \int h (R_\varepsilon - 1) (t, x) dx + \frac{1}{2} \int R_\varepsilon |U_\varepsilon - \varepsilon \tilde{u}|^2 (t, x) dx$$
$$+ \frac{3}{2} \int R_\varepsilon (T_\varepsilon - \log T_\varepsilon - 1) (t, x) dx$$

The difficulty to apply this strategy is to obtain a control on the relaxation to local Maxwellians. Indeed, in the case of the Boltzmann equation, the entropy production is not known to measure the distance between f_ε and M_{f_ε} in some suitable sense (see Chapter 3 for a brief discussion on that point).

• The first step consists therefore in introducing a suitable decomposition of the flux term (as in the case of the Navier-Stokes limit), well adapted to the structure of the collision operator :

$$\frac{1}{\varepsilon^2} \int \Phi_\varepsilon(f_\varepsilon - M_\varepsilon)(t, x, v) dv = \int \Phi_\varepsilon M_\varepsilon \left(\frac{1}{\varepsilon} \tilde{g}_\varepsilon + \frac{1}{4} \tilde{g}_\varepsilon^2\right)(t, x, v) dv$$

$$= \int \tilde{\Phi}_\varepsilon M_\varepsilon \frac{1}{\varepsilon} \mathcal{L}_\varepsilon \tilde{g}_\varepsilon(t, x, v) dv + \frac{1}{4} \int \Phi_\varepsilon M_\varepsilon \tilde{g}_\varepsilon^2(t, x, v) dv$$

where $\tilde{\Phi}_\varepsilon$ is the pseudo-inverse by $\mathcal{L}_\varepsilon$ of Φ_ε, which can be obtained from $\tilde{\Phi}$ defined in Remark 3.2.3 by translation.

Then, using the identity

$$\frac{1}{\varepsilon}\mathcal{M}_\varepsilon\mathcal{L}_\varepsilon\tilde{g}_\varepsilon = -\frac{2}{\varepsilon^2}Q(\sqrt{\mathcal{M}_\varepsilon f_\varepsilon}, \sqrt{\mathcal{M}_\varepsilon f_\varepsilon}) + \frac{1}{2}Q(\mathcal{M}_\varepsilon\tilde{g}_\varepsilon, \mathcal{M}_\varepsilon\tilde{g}_\varepsilon) \qquad (5.37)$$

which is the translated variant of (3.19) used in Chapter 3, we eventually arrive at the following decomposition

$$\begin{aligned}
\frac{1}{\varepsilon^2}\int \Phi_\varepsilon(f_\varepsilon - \mathcal{M}_\varepsilon)(t,x,v)dv &= -2\varepsilon^{(q-1)/2}\int \tilde{\Phi}_\varepsilon\mathcal{M}_\varepsilon\tilde{q}_\varepsilon(t,x,v)dv \\
&+\frac{1}{2}\int \tilde{\Phi}_\varepsilon Q(\mathcal{M}_\varepsilon\tilde{g}_\varepsilon, \mathcal{M}_\varepsilon\tilde{g}_\varepsilon)(t,x,v)dv \\
&+\frac{1}{4}\int \Phi_\varepsilon\mathcal{M}_\varepsilon\tilde{g}_\varepsilon^2(t,x,v)dv \\
&\overset{\text{def}}{=} I_1 + I_2 + I_3
\end{aligned} \qquad (5.38)$$

The term I_1 measures in some sense the relaxation of f_ε to the manifold of local Maxwellians, and is controlled by (5.24) :

$$\|I_1\|_{L^2(dtdx)} \le 2\varepsilon^{(q-1)/2}\|\tilde{q}_\varepsilon\|_{L^2(dtdx\nu_\varepsilon^{-1}\mathcal{M}_\varepsilon dv)}\|\tilde{\Phi}_\varepsilon\|_{L^2(\nu_\varepsilon\mathcal{M}_\varepsilon dv)} \, ,$$

from which we conclude that

$$I_1 = o(1)_{L^2(dxdt)} \qquad (5.39)$$

Estimating the term I_2 requires to obtain some further integrability (with respect to v) on $\tilde{g}_\varepsilon$. Indeed, using the continuity of the collision operator, namely

$$\left\|\frac{1}{\mathcal{M}_\varepsilon}Q(\mathcal{M}_\varepsilon\tilde{g}, \mathcal{M}_\varepsilon\tilde{g})\right\|_{L^2(\mathcal{M}_\varepsilon\nu_\varepsilon^{-1}dv)} \le C\|\tilde{g}\|_{L^2(\mathcal{M}_\varepsilon dv)}\|\tilde{g}\|_{L^2(\mathcal{M}_\varepsilon(1+|v|)^\beta dv)},$$

we have

$$\|I_2(t)\|_{L^1(dx)} \le C\|\tilde{\Phi}_\varepsilon\|_{L^2(\nu_\varepsilon\mathcal{M}_\varepsilon dv)}\|(\tilde{g}_\varepsilon)^2(t)\|_{L^1(dx(1+|v|)^\beta\mathcal{M}_\varepsilon dv)}.$$

Controlling the term I_3 requires still further integrability (with respect to v) since $\Phi_\varepsilon = O(|v|^2)$ as $|v| \to \infty$.

$$\|I_3(t)\|_{L^1(dx)} \le C\|(\tilde{g}_\varepsilon)^2(t)\|_{L^1(dx(1+|v|)^2\mathcal{M}_\varepsilon dv)}.$$

• We will obtain this additional integrability using the same arguments as in Lemma 3.2.5, namely the relaxation estimate (5.25)

$$\|\tilde{g}_\varepsilon - \Pi_\varepsilon\tilde{g}_\varepsilon\|_{L^2(\mathcal{M}_\varepsilon dv)} = O(\varepsilon^{(q+1)/2})_{L^2_{t,x}} + O(\varepsilon)\|\tilde{g}_\varepsilon\|^2_{L^2(\mathcal{M}_\varepsilon dv)},$$

together with Young's inequality. The crucial point is of course the fact that, by definition of $\Pi_\varepsilon\tilde{g}_\varepsilon$, for all $p < +\infty$

$$(\Pi_\varepsilon \tilde{g}_\varepsilon)^2 = O\left(\frac{1}{\varepsilon^2} H(f_\varepsilon | \mathcal{M}_\varepsilon)\right)_{L^1(dx, L^p(\mathcal{M}_\varepsilon dv))}.$$

We first introduce some truncation

$$\tilde{\gamma}_\varepsilon = \gamma\left(\frac{f_\varepsilon}{\mathcal{M}_\varepsilon}\right)$$

where $\gamma \in C_c^\infty(\mathbf{R}^+, [0,1])$ satisfies $\gamma_{|[0,2]} \equiv 1$, so that

$$\varepsilon \tilde{\gamma}_\varepsilon \tilde{g}_\varepsilon = O(1)_{L_{t,x,v}^\infty}, \quad \|\tilde{\gamma}_\varepsilon \tilde{g}_\varepsilon\|_{L^2(dx \mathcal{M}_\varepsilon dv)} = O\left(\frac{1}{\varepsilon} H(f_\varepsilon | \mathcal{M}_\varepsilon)^{1/2}\right).$$

For moderated tails, we have

$$\tilde{\gamma}_\varepsilon \tilde{g}_\varepsilon^2 = \varepsilon \tilde{\gamma}_\varepsilon \tilde{g}_\varepsilon \left(\frac{\tilde{g}_\varepsilon - \Pi_\varepsilon \tilde{g}_\varepsilon}{\varepsilon}\right) + \tilde{\gamma}_\varepsilon \tilde{g}_\varepsilon \Pi_\varepsilon \tilde{g}_\varepsilon$$

$$= O\left(\frac{1}{\varepsilon^2} H(f_\varepsilon | \mathcal{M}_\varepsilon)\right)_{L_t^\infty(L_x^1(L^p(\mathcal{M}_\varepsilon dv)))} + o(1)_{L^2(dt dx \mathcal{M}_\varepsilon dv)}$$

for all $p < 2$. In particular

$$\tilde{\gamma}_\varepsilon \tilde{g}_\varepsilon^2(1 + |v|^2) = O\left(\frac{1}{\varepsilon^2} H(f_\varepsilon | \mathcal{M}_\varepsilon)\right)_{L_t^\infty(L^1(dx \mathcal{M}_\varepsilon dv))} + o(1)_{L^2(dt dx, L^1(\mathcal{M}_\varepsilon dv))}$$

$$(5.40)$$

On the other hand, by Young's inequality, we have

$$(1 + |v|)^2 \tilde{g}_\varepsilon^2(1 - \tilde{\gamma}_\varepsilon) \leq \frac{10}{\varepsilon^2} \left|\frac{f_\varepsilon}{\mathcal{M}_\varepsilon} - 1\right| \frac{(1 + |v|)^2}{10}(1 - \tilde{\gamma}_\varepsilon)$$

$$\leq \frac{10}{\varepsilon^2} h\left(\frac{f_\varepsilon}{\mathcal{M}_\varepsilon} - 1\right) + \frac{10}{\varepsilon^2}(1 - \tilde{\gamma}_\varepsilon)^2 e^{\frac{(1+|v|)^2}{10}} \qquad (5.41)$$

The first term in the right-hand side is controlled by the modulated entropy, whereas the second one is dealt with using again the relaxation estimate. We have indeed

$$(1 - \tilde{\gamma}_\varepsilon) = O(1)_{L_{t,x,v}^\infty}, \quad \left\|\frac{1 - \tilde{\gamma}_\varepsilon}{\varepsilon}\right\|_{L^2(dx \mathcal{M}_\varepsilon dv)} = O\left(\frac{1}{\varepsilon} H(f_\varepsilon | \mathcal{M}_\varepsilon)^{1/2}\right),$$

and

$$\frac{1 - \tilde{\gamma}_\varepsilon}{\varepsilon} \leq (1 - \tilde{\gamma}_\varepsilon)\tilde{g}_\varepsilon \leq (1 - \tilde{\gamma}_\varepsilon)(\tilde{g}_\varepsilon - \Pi_\varepsilon \tilde{g}_\varepsilon) + (1 - \tilde{\gamma}_\varepsilon)\Pi_\varepsilon \tilde{g}_\varepsilon$$

$$= O\left(\frac{1}{\varepsilon} H(f_\varepsilon | \mathcal{M}_\varepsilon)\right)_{L^1(dx, L^2(\mathcal{M}_\varepsilon dv))} + o(\varepsilon)_{L^2(dt dx \mathcal{M}_\varepsilon dv)}$$

$$+ O\left(\frac{1}{\varepsilon^2} H(f_\varepsilon | \mathcal{M}_\varepsilon)\right)_{L^2(dx, L^p(\mathcal{M}_\varepsilon dv))}$$

for any $p < +\infty$, from which we deduce that

$$\left(\frac{1 - \tilde{\gamma}_\varepsilon}{\varepsilon}\right)^2 e^{\frac{(1+|v|)^2}{10}} = O\left(\frac{1}{\varepsilon^2} H(f_\varepsilon | \mathcal{M}_\varepsilon)\right)_{L_t^\infty(L^1(dx \mathcal{M}_\varepsilon dv))} + o(1)_{L^2(dt dx, L^1(\mathcal{M}_\varepsilon dv))}$$

$$(5.42)$$

- By (5.40), (5.41) and (5.42), we get finally

$$|I_2| + |I_3| = O\left(\frac{1}{\varepsilon^2}H(f_\varepsilon|\mathcal{M}_\varepsilon)\right)_{L^1(dx)} + o(1)_{L^2(dtdx)} \tag{5.43}$$

Combining (5.38) with (5.39) and (5.43) leads then to

$$\frac{1}{\varepsilon^2}\int \Phi_\varepsilon(f_\varepsilon - \mathcal{M}_\varepsilon)(t, x, v)dv = O\left(\frac{1}{\varepsilon^2}H(f_\varepsilon|\mathcal{M}_\varepsilon)\right)_{L^1(dx)} + o(1)_{L^2(dtdx)}$$

which completes the proof of Proposition 5.3.1. $\qquad\qquad\square$

5.3.2 Proof of Theorem 5.1.8 and Corollary 5.1.9

We deduce from (5.26) and (5.36) that

$$\frac{1}{\varepsilon^2}H(f_\varepsilon|\mathcal{M}_\varepsilon)(t) + \frac{1}{\varepsilon^2}\int_{\mathbf{R}^3} \mathrm{tr}(m_\varepsilon)(t) + \frac{1}{\varepsilon^{q+3}}\int_0^t\int D(f_\varepsilon)dxds$$
$$\leq \frac{1}{\varepsilon^2}H(f_{\varepsilon,in}|\mathcal{M}_{\varepsilon,in}) + o(1)$$
$$+ C\int_0^t \|\nabla_x\tilde{u} + (\nabla_x\tilde{u})^T\|_{L^2\cap L^\infty}\left(\frac{1}{\varepsilon^2}H(f_\varepsilon|\mathcal{M}_\varepsilon) + \frac{1}{\varepsilon^2}\int_\Omega \mathrm{tr}(m_\varepsilon)\right)(s)ds$$
$$+ \frac{1}{\varepsilon}\int_0^t\int A(\tilde{u})\cdot\int(\varepsilon\tilde{u} - v)f_\varepsilon(s, x, v)dvdxds$$

Integrating next this differential inequality leads to

$$\frac{1}{\varepsilon^2}H(f_\varepsilon|\mathcal{M}_\varepsilon)(t) + \frac{1}{\varepsilon^2}\int_\Omega \mathrm{tr}(m_\varepsilon)(t) + \frac{1}{\varepsilon^{q+3}}\int_0^t\int D(f_\varepsilon)dxds$$
$$\leq \frac{1}{\varepsilon^2}H(f_{\varepsilon,in}|\mathcal{M}_{\varepsilon,in})\exp\left(C\int_0^t \|\nabla_x\tilde{u} + (\nabla_x\tilde{u})^T\|_{L^2\cap L^\infty}(s)ds\right) + o(1)$$
$$+ \int_0^t \exp\left(C\int_s^t \|\nabla_x\tilde{u} + (\nabla_x\tilde{u})^T\|_{L^2\cap L^\infty}(\tau)d\tau\right)\iint A(\tilde{u})\cdot(\tilde{u} - \frac{v}{\varepsilon})f_\varepsilon dvdxds \tag{5.44}$$

for any divergence free vector-field $\tilde{u} \in C_c^\infty(\mathbf{R}^+\times\Omega)$ such that $n\cdot\tilde{u}_{|\partial\Omega} = 0$.

- In the general case, we will then deduce that any limit point of the sequence of scaled bulk velocities (u_ε) is a dissipative solution to the incompressible Euler equations (5.1)(5.3).

Let us first recall that, by Lemma 3.1.2 in Chapter 3, up to extraction of a subsequence, we have the convergences

$$u_\varepsilon = \frac{1}{\varepsilon}\int f_\varepsilon vdv \rightharpoonup u \text{ weakly in } L^1_{loc}(dtdx),$$

$$R_\varepsilon = \int f_\varepsilon dv \to 1 \text{ strongly in } L^1_{loc}(dtdx).$$

Furthermore, taking limits in the local conservation of mass

$$\partial_t \int f_\varepsilon dv + \frac{1}{\varepsilon} \nabla_x \cdot \int f_\varepsilon v dv = 0$$

shows that

$$\nabla_x \cdot u = 0.$$

In order to prove that u is a dissipative solution to the incompressible Euler equations (5.1)(5.3), it remains then to check that it satisfies the stability inequality (5.11). Denote as previously by $\mathcal{M}_{f_\varepsilon}$ the local Maxwellian having the same moments as f_ε. Simple computations give

$$\frac{1}{\varepsilon^2} H(f_\varepsilon|\mathcal{M}_\varepsilon) = \frac{1}{\varepsilon^2} H(\mathcal{M}_{f_\varepsilon}|\mathcal{M}_\varepsilon) + \frac{1}{\varepsilon^2} H(f_\varepsilon|\mathcal{M}_{f_\varepsilon})$$
$$\geq \frac{1}{\varepsilon^2} H(\mathcal{M}_{f_\varepsilon}|\mathcal{M}_\varepsilon) \geq \frac{1}{2} \int \frac{(u_\varepsilon - R_\varepsilon \tilde{u})^2}{R_\varepsilon} dx,$$

By convexity of the functional

$$(R, u) \mapsto \frac{(u - R\tilde{u})^2}{R},$$

since $R_\varepsilon \to 1$ and $u_\varepsilon \to u$ in the vague sense of measures,

$$\frac{1}{2} \|u - \tilde{u}\|^2_{L^2(\Omega)} \leq \liminf_{\varepsilon \to 0} \frac{1}{2} \int \frac{(u_\varepsilon - R_\varepsilon \tilde{u})^2}{R_\varepsilon} dx.$$

Combining these inequalities with (5.44) leads then to

$$\frac{1}{2} \|(u - \tilde{u})(t)\|^2_{L^2(\Omega)}$$
$$\leq \left(\liminf_{\varepsilon \to 0} \frac{1}{\varepsilon^2} H(f_{\varepsilon,in}|\mathcal{M}_{\varepsilon,in}) \right) \exp\left(C \int_0^t \|\nabla_x \tilde{u} + (\nabla_x \tilde{u})^T\|_{L^2 \cap L^\infty}(s)ds \right)$$
$$+ \int_0^t \exp\left(C \int_s^t \|\nabla_x \tilde{u} + (\nabla_x \tilde{u})^T\|_{L^2 \cap L^\infty}(\tau)d\tau \right) \int A(\tilde{u}) \cdot (\tilde{u} - u)(s)dxds.$$

Then, from (5.16) and the identity

$$\frac{1}{\varepsilon^2} H(f_{\varepsilon,in}|\mathcal{M}_{\varepsilon,in}) = \frac{1}{\varepsilon^2} H(f_{\varepsilon,in}|\mathcal{M}_{(1,\varepsilon u_{in},1)}) + \frac{1}{2} \iint f_{\varepsilon,in}(u_{in} - \tilde{u}_{in})^2 dvdx$$
$$+ \frac{1}{\varepsilon} \iint f_{\varepsilon,in}(v - \varepsilon u_{in}) \cdot (u_{in} - \tilde{u}_{in})dvdx$$
$$= \frac{1}{\varepsilon^2} H(f_{\varepsilon,in}|\mathcal{M}_{(1,\varepsilon u_{in},1)}) + \frac{1}{2} \|u_{in} - \tilde{u}_{in}\|^2_{L^2(\Omega)}$$
$$+ \frac{1}{2} \iint (f_{\varepsilon,in} - \mathcal{M}_{(1,\varepsilon u_{in},1)})(u_{in} - \tilde{u}_{in})^2 dvdx$$
$$+ \frac{1}{\varepsilon} \iint (f_{\varepsilon,in} - \mathcal{M}_{(1,\varepsilon u_{in},1)})(v - \varepsilon u_{in}) \cdot (u_{in} - \tilde{u}_{in})dvdx$$

we deduce, using Young's inequality, that

$$\frac{1}{\varepsilon^2} H(f_{\varepsilon,in}|\mathcal{M}_{(1,\varepsilon\tilde{u}_{in},1)}) = \frac{1}{2}\|u_{in} - \tilde{u}_{in}\|^2_{L^2(\Omega)} + o(1).$$

Therefore

$$\frac{1}{2}\|(u - \tilde{u})(t)\|^2_{L^2(\Omega)}$$
$$\leq \frac{1}{2}\|u_{in} - \tilde{u}_{in}\|^2_{L^2(\Omega)} \exp\left(C \int_0^t \|\nabla_x\tilde{u} + (\nabla_x\tilde{u})^T\|_{L^2\cap L^\infty}(s)ds\right)$$
$$+ \int_0^t \exp\left(C \int_s^t \|\nabla_x\tilde{u} + (\nabla_x\tilde{u})^T\|_{L^2\cap L^\infty}(\tau)d\tau\right) \int A(\tilde{u}) \cdot (\tilde{u} - u)(s)dxds.$$

for any divergence free vector-field $\tilde{u} \in C_c^\infty(\mathbf{R}^+ \times \Omega)$ such that $n \cdot \tilde{u}_{|\partial\Omega} = 0$.

By a standard density argument, the previous inequality can be extended to any divergence free vector-field

$$\tilde{u} \in L^\infty([0, t^*], L^2(\Omega)) \cap L^1([0, t^*], H^1 \cap W^{1,\infty}(\Omega)))$$

such that $n \cdot \tilde{u}_{|\partial\Omega} = 0$. We conclude that u is a dissipative solution to (5.1)(5.3).

• In the case where the incompressible Euler equations (5.1)(5.3) have a unique smooth solution u on $[0, t^*]$ with initial data u_{in}, we know of course that any dissipative solution of (5.1)(5.3) will coincide with u. In particular, the sequence of scaled bulk velocities (u_ε) converge to u. We have furthermore the entropic convergence of the sequence $(g_\varepsilon(t))$ defined by $f_\varepsilon = M(1 + \varepsilon g_\varepsilon)$ for all $t \in [0, t^*]$.

Indeed, by the assumption (5.16) on the initial data, we see that

$$\frac{1}{\varepsilon^2} H(f_\varepsilon|\mathcal{M}_{(1,\varepsilon u,1)})(t) \to 0 \text{ as } \varepsilon \to 0.$$

5.4 Taking into Account Acoustic Waves

The proof of Theorem 5.1.11 requires some improvements of the relative entropy method developed previously. The main idea is that, in domains where the distribution is expected to present rapid variations, the formal hydrodynamic approximation is not relevant, and that correctors have to be added in order to obtain the convenient asymptotics. The point is indeed to obtain a refined decription of the asymptotics taking into account both the relaxation in the initial layer and the acoustic waves.

Taking into account acoustic waves (and the heat equation) does not modify strongly the method since they only contribute to the hydrodynamic part of the distribution. We indeed recall from Section 5.2 that, under the additional assumption (5.19), for any $(\tilde{\rho}, \tilde{u}, \tilde{\theta}) \in C_c^\infty(\mathbf{R}^+ \times \Omega)$, denoting

$$\mathcal{M}_\varepsilon \overset{\text{def}}{=} \mathcal{M}_{(\exp(\varepsilon\tilde\rho),\varepsilon\tilde u,\exp(\varepsilon\tilde\theta))},$$

we have the following modulated entropy inequality

$$H\left(f_\varepsilon|\mathcal{M}_\varepsilon\right)(t) + \frac{1}{\varepsilon^{q+1}} \int_0^t \int D(f_\varepsilon)\,ds\,dx$$

$$\leq H(f_{\varepsilon,in}|\mathcal{M}_{\varepsilon,in}) + \int_0^t \int \partial_t \exp(\varepsilon\tilde\rho)\,dx\,ds$$

$$-\varepsilon \int_0^t \iint f_\varepsilon\left(1, e^{-\varepsilon\tilde\theta}(v-\varepsilon\tilde u), \frac{1}{2}\left(e^{-\varepsilon\tilde\theta}|v-\varepsilon\tilde u|^2 - 3\right)\right)\cdot A_\varepsilon(\tilde\rho,\tilde u,\tilde\theta)\,dv\,dx\,ds$$

$$-\int_0^t \iint f_\varepsilon \nabla_x \tilde u : \Phi_\varepsilon\,dx\,dv + \iint f_\varepsilon e^{\frac{1}{2}\varepsilon\tilde\theta} \nabla_x\tilde\theta \cdot \Psi_\varepsilon\,dx\,dv\,ds$$

for any solution to the Boltzmann equation satisfying the conservation laws.

5.4.1 Construction of Approximate Solutions by a Filtering Method

We then have to construct a sequence of approximate solutions to the systems

$$\mathbf{A}_\varepsilon(\rho, u, \theta) = 0,$$

or in other words to the systems

$$\partial_t \rho + (u\cdot\nabla_x)\rho + \frac{1}{\varepsilon}\nabla_x\cdot u = 0,$$

$$\partial_t u + (u\cdot\nabla_x)u + \left(\frac{e^{\varepsilon\theta}-1}{\varepsilon}\right)\nabla_x(\rho+\theta) + \frac{1}{\varepsilon}\nabla_x(\rho+\theta) = 0, \qquad (5.45)$$

$$\partial_t \theta + (u\cdot\nabla_x)\theta + \frac{2}{3\varepsilon}\nabla_x\cdot u = 0$$

More precisely, we will require that

$$\mathbf{A}_\varepsilon(\rho_\varepsilon, u_\varepsilon, \theta_\varepsilon) \to 0 \text{ in } L^2(dt\,dx) \text{ as } \varepsilon \to 0. \qquad (5.46)$$

One of the difficulty here (in comparison with classical penalization problems) is that we further need that these approximate solutions conserve the total mass at higher order

$$\frac{1}{\varepsilon^2}\int \partial_t \exp(\varepsilon\rho_\varepsilon)\,dx \to 0 \text{ in } L^1(dt) \text{ as } \varepsilon \to 0. \qquad (5.47)$$

(Note that, for exact solutions, the total mass is exactly conserved.)

Such a construction is done by a *filtering method* (see [60] or [96] for instance), i.e. considering the family $\mathcal{W}\left(\frac{t}{\varepsilon}\right)(\rho, u, \theta)$ where $\mathcal{W}$ is the semigroup generated by the linear penalization operator W defined by

$$W(\rho, u, \theta) = \left(\nabla_x\cdot u, \nabla_x(\rho+\theta), \frac{2}{3}\nabla_x\cdot u\right).$$

The first order approximation is then obtained by taking (strong) limits in the filtered system. Nevertheless, because of the high frequency oscillations, we do not expect the error in this first order approximation to converge strongly to 0.

We therefore have to add some correctors (i.e. the second and third order approximations) in order to establish the convergence statements (5.46)(5.47).

More precisely, we have the following

Proposition 5.4.1 *Let* $(\rho_{in}, u_{in}, \theta_{in})$ *belong to* $H^s(\Omega)$ *for some* $s > \frac{5}{2}$. *Then there exist some* $t^* > 0$, *and some family* $(\rho_\varepsilon^N, u_\varepsilon^N, \theta_\varepsilon^N)$ *satisfying the uniform bound*

$$\sup_{N \in \mathbf{N}} \lim_{\varepsilon \to 0} \|(\rho_\varepsilon^N, u_\varepsilon^N, \theta_\varepsilon^N)\|_{L^1([0,t^*], H^s(\Omega))} \leq C \qquad (5.48)$$

and such that the following convergences hold as $\varepsilon \to 0$ *then* $N \to \infty$:

$$(\rho_{\varepsilon,in}^N, u_{\varepsilon,in}^N, \theta_{\varepsilon,in}^N) \to (\rho_{in}, u_{in}, \theta_{in}) \text{ in } H^s(dx), \qquad (5.49)$$

$$\mathbf{A}_\varepsilon(\rho_\varepsilon^N, u_\varepsilon^N, \theta_\varepsilon^N) \to 0 \text{ in } L^2(dtdx), \qquad (5.50)$$

$$\frac{1}{\varepsilon^2} \int \partial_t \exp(\varepsilon \rho_\varepsilon^N) dx \to 0 \text{ in } L^1(dt). \qquad (5.51)$$

Proof. Since the proof of that proposition is very technical, we will only sketch the main arguments here, and refer to [95] for the details.

Let us first introduce some notations to recast the system

$$\mathbf{A}_\varepsilon(\rho, u, \theta) = 0$$

in a suitable form. For any $V = (\rho, u, \theta)$ we define the symmetric bilinear form $\mathcal{B}$ by

$$\mathcal{B}(V, V) = \begin{pmatrix} (u \cdot \nabla_x)\rho \\ (u \cdot \nabla_x)u + \theta \nabla_x(\rho + \theta) \\ (u \cdot \nabla_x)\theta \end{pmatrix}$$

We are therefore interested in the (approximate) solutions to

$$\partial_t V + \frac{1}{\varepsilon} W V + \mathcal{B}(V, V) = - \begin{pmatrix} 0 \\ \frac{1}{\varepsilon}(e^{\varepsilon\theta} - 1 - \varepsilon\theta)\nabla_x(\rho + \theta) \\ 0 \end{pmatrix}$$

which are also approximate solutions (in the sense of (5.50)) to

$$\partial_t V + \frac{1}{\varepsilon} W V + \mathcal{B}(V, V) = 0$$

provided that V is uniformly bounded in $L^\infty([0, t^*], W^{1,\infty} \cap L^2(\Omega))$. Let us also recall that we further need that these approximate solutions satisfy some global conservation of mass (5.51).

- We first conjugate the system by the semi-group $\mathcal{W}\left(\frac{t}{\varepsilon}\right)$ generated by W

$$\partial_t\left(\mathcal{W}\left(\frac{t}{\varepsilon}\right)V\right) + \mathcal{W}\left(\frac{t}{\varepsilon}\right)\mathcal{B}(V,V) = 0,$$

or equivalently

$$\partial_t\tilde{V} + \mathcal{W}\left(\frac{t}{\varepsilon}\right)\mathcal{B}\left(\mathcal{W}\left(-\frac{t}{\varepsilon}\right)\tilde{V}, \mathcal{W}\left(-\frac{t}{\varepsilon}\right)\tilde{V}\right) = 0 \qquad (5.52)$$

denoting by $\tilde{V}$ the filtered field $\tilde{V} = \mathcal{W}\left(\frac{t}{\varepsilon}\right)V$.

We therefore expect the solutions (and approximate solutions) to (5.52) to have a very different behaviour depending on the nature of the spectrum of W. In the case when Ω is a smooth bounded domain, $(Id - \Delta)^{-1}$ is a compact operator with discrete spectrum, from which we deduce that W has discrete spectrum. The (formal) limit system depends therefore on the resonances between acoustic modes. In the case when Ω is an exterior domain, the Laplacian has continuous spectrum and one can prove using dispersion properties that the corresponding acoustic waves converge strongly to 0.

For the sake of simplicity, we will focus here on the case of bounded domains (or of the torus $\mathbf{T}^3$). We will denote by $(i\lambda_k)$ the sequence of eigenvalues of W corresponding to the boundary condition of Neumann type

$$u \cdot n = 0 \text{ on } \partial\Omega,$$

and by Π_λ the projection on $\mathrm{Ker}(W - \lambda Id)$.

- At leading order, we then obtain

$$\partial_t\tilde{V}_0 + \mathcal{B}_W(\tilde{V}_0, \tilde{V}_0) = 0. \qquad (5.53)$$

denoting by $\mathcal{B}_W$ the limiting quadratic operator defined by

$$\mathcal{B}_W = \sum_k \sum_{\lambda_{k_1}+\lambda_{k_2}=\lambda_k} \Pi_{\lambda_k}\mathcal{B}(\Pi_{\lambda_{k_1}}\cdot, \Pi_{\lambda_{k_2}}\cdot). \qquad (5.54)$$

An algebraic computation (which is the basic argument in the compensated compactness method, see [77] or Proposition 4.3.2 in the previous chapter) shows that, for all $\lambda, \mu \neq 0$

$$\Pi_0\mathcal{B}(\Pi_\lambda\tilde{V}_0, \Pi_\mu\tilde{V}_0) = 0.$$

Indeed we have the following formula for Π_0

$$\Pi_0(\rho, u, \theta) = \left(\frac{2\rho - 3\theta}{5}, Pu, \frac{3\theta - 2\rho}{5}\right).$$

Then, with the notations $\Pi_\lambda \tilde{V}_0 = (\rho_\lambda, u_\lambda, \theta_\lambda)$ and $\Pi_\mu \tilde{V}_0 = (\rho_\mu, u_\mu, \theta_\mu)$, we get

$$\Pi_0 \mathcal{B}(\Pi_\lambda \tilde{V}_0, \Pi_\mu \tilde{V}_0) = \frac{1}{2} \Pi_0 \begin{pmatrix} (u_\lambda \cdot \nabla_x)\rho_\mu + (u_\mu \cdot \nabla_x)\rho_\lambda \\ (u_\lambda \cdot \nabla_x)u_\mu + (u_\mu \cdot \nabla_x)u_\lambda + \frac{5}{2}\theta_\mu \nabla_x \theta_\lambda + \frac{5}{2}\theta_\lambda \nabla_x \theta_\mu \\ (u_\lambda \cdot \nabla_x)\theta_\mu + (u_\mu \cdot \nabla_x)\theta_\lambda \end{pmatrix}$$
$$= \frac{1}{10} \begin{pmatrix} (u_\lambda \cdot \nabla_x)(2\rho_\mu - 3\theta_\mu) + (u_\mu \cdot \nabla_x)(2\rho_\lambda - 3\theta_\lambda) \\ 5P\left((u_\lambda \cdot \nabla_x)u_\mu + (u_\mu \cdot \nabla_x)u_\lambda\right) \\ (u_\lambda \cdot \nabla_x)(3\theta_\mu - 2\rho_\mu) + (u_\mu \cdot \nabla_x)(3\theta_\lambda - 2\rho_\lambda) \end{pmatrix}$$
$$= \frac{1}{2} \begin{pmatrix} 0 \\ P\left(\nabla_x(u_\lambda \cdot u_\mu) - u_\mu \wedge (\nabla_x \wedge u_\lambda) - u_\lambda \wedge (\nabla_x \wedge u_\mu)\right) \\ 0 \end{pmatrix}$$

$$(5.55)$$

since $\nabla_x \wedge u_\lambda = 0$ and $3\theta_\lambda - 2\rho_\lambda = 0$.

In other words the equation governing the non-oscillating part can be decoupled from the rest of the system

$$\partial_t \Pi_0 \tilde{V}_0 + \Pi_0 \mathcal{B}(\Pi_0 \tilde{V}_0, \Pi_0 \tilde{V}_0) = 0,$$

which can be rewritten

$$\begin{aligned} \partial_t \bar{\rho} + (\bar{u} \cdot \nabla_x)\bar{\rho} = 0, \quad \nabla_x(\bar{\rho} + \bar{\theta}) = 0, \\ \partial_t \bar{u} + (\bar{u} \cdot \nabla_x)\bar{u} + \nabla_x p = 0, \quad \nabla_x \cdot \bar{u} = 0, \end{aligned} \qquad (5.56)$$

where $(\bar{\rho}, \bar{u}, \bar{\theta}) = \Pi_0 \tilde{V}_0 = \Pi_0 V_0$. Note in particular that

$$\partial_t \int \bar{\rho} \, dx = 0.$$

A classical study based on harmonic analysis allows to prove that (5.53) has a unique strong solution $V_0 \in L^\infty_{loc}([0, t^*), H^s(\Omega))$ provided that $V_{in} \in H^s(\Omega)$ for $s > \frac{5}{2}$. The point is to check that

$$\|V\|^2_{H^s(\Omega)} \sim \|\Pi_0 V\|^2_{H^s(\Omega)} + \sum_k (1 + \lambda_k^2)^s \|\Pi_k V\|^2_{L^2(\Omega)}$$

which comes from the fact that the acoustic operator acts as a derivation on the orthogonal complement of its kernel.

For that solution, we also have a uniform bound on the time derivative

$$\partial_t \tilde{V}_0 \in L^\infty_{loc}([0, t^*), H^{s-1}(\Omega)).$$

Remarking that, for all $\lambda \neq 0$,

$$\int \rho_\lambda \, dx = \frac{1}{i\lambda} \int (\nabla_x \cdot u_\lambda) \, dx = \frac{1}{i\lambda} \int_{\partial\Omega} u_\lambda \cdot n \, d\sigma_x = 0, \qquad (5.57)$$

we get, denoting $(\rho_0, u_0, \theta_0) = V_0 = \mathcal{W}\left(-\frac{t}{\varepsilon}\right) \tilde{V}_0$,

$$\partial_t \int \rho_0 \, dx = \partial_t \int \bar{\rho} \, dx = 0.$$

Furthermore, denoting by J_N the projection on the N first eigenmodes,

$$\tilde{V}_0 - J_N \tilde{V}_0 \to 0 \text{ as } N \to \infty \text{ in } L^\infty_{loc}([0,t^*), H^s(\Omega)) \cap W^{1,\infty}_{loc}([0,t^*), H^{s-1}(\Omega)),$$

and

$$\int \rho_0^N \, dx = \int \rho_0 dx.$$

Note however that $\tilde{V}_0$ (and consequently $\tilde{V}_0^N = J_N \tilde{V}_0$) is not an approximate solution to (5.52) in the sense of (5.50). We have indeed

$$\partial_t \tilde{V}_0^N + \mathcal{W}\left(\frac{t}{\varepsilon}\right) \mathcal{B}\left(\mathcal{W}\left(-\frac{t}{\varepsilon}\right)\tilde{V}_0^N, \mathcal{W}\left(-\frac{t}{\varepsilon}\right)\tilde{V}_0^N\right)$$
$$= (Id - J_N)\mathcal{B}_W(\tilde{V}_0, \tilde{V}_0) + \mathcal{B}_W(\tilde{V}_0^N - \tilde{V}_0, \tilde{V}_0^N + \tilde{V}_0)$$
$$+ \mathcal{W}\left(\frac{t}{\varepsilon}\right)\mathcal{B}\left(\mathcal{W}\left(-\frac{t}{\varepsilon}\right)\tilde{V}_0^N, \mathcal{W}\left(-\frac{t}{\varepsilon}\right)\tilde{V}_0^N\right) - \mathcal{B}_W(\tilde{V}_0^N, \tilde{V}_0^N)$$

where the last term is expected to be an oscillating term, that is to converge weakly but not strongly to 0. We therefore need to add some correctors.

• The second order approximation V_1^N is defined by

$$\tilde{V}_1^N = J_N \sum_{\lambda_{k_1} + \lambda_{k_2} \neq \lambda_k} \frac{\exp\left(\frac{it}{\varepsilon}(\lambda_k - \lambda_{k_1} - \lambda_{k_2})\right)}{i(\lambda_{k_1} + \lambda_{k_2} - \lambda_k)} \Pi_{\lambda_k}\mathcal{B}(\Pi_{\lambda_{k_1}}\tilde{V}_0^N, \Pi_{\lambda_{k_2}}\tilde{V}_0^N)$$

From the bounds on $\tilde{V}_0$ (which are clearly inherited by $\tilde{V}_0^N$) and the definition of $\mathcal{B}$, we deduce that, for all $N > 0$ and all $\sigma > 0$

$$\tilde{V}_1^N \text{ in } L^\infty_{loc}([0,t^*), H^\sigma(\Omega))$$

(with an estimate depending on N), and that, for all $t < t^*$

$$\sup_{N \in \mathbb{N}} \lim_{\varepsilon \to 0} \|\tilde{V}_0^N + \varepsilon \tilde{V}_1^N\|_{L^\infty([0,t], H^s(\Omega))} \leq C\|\tilde{V}_0\|_{L^\infty([0,t^*], H^s(\Omega))}.$$

It remains then to check that $\tilde{V}_0^N + \varepsilon \tilde{V}_1^N$ is an approximate solution to (5.52) in a strong sense.

$$\partial_t(\tilde{V}_0^N + \varepsilon\tilde{V}_1^N) + \mathcal{W}\left(\frac{t}{\varepsilon}\right)\mathcal{B}\left(\mathcal{W}\left(\frac{-t}{\varepsilon}\right)(\tilde{V}_0^N + \varepsilon\tilde{V}_1^N), \mathcal{W}\left(\frac{-t}{\varepsilon}\right)(\tilde{V}_0^N + \varepsilon\tilde{V}_1^N)\right)$$
$$= (Id - J_N)\mathcal{B}_W(\tilde{V}_0, \tilde{V}_0) + \mathcal{B}_W(\tilde{V}_0^N - \tilde{V}_0, \tilde{V}_0 + \tilde{V}_0^N)$$
$$+ (Id - J_N) \sum_{\lambda_{k_1} + \lambda_{k_2} \neq \lambda_k} \exp\left(i\frac{t}{\varepsilon}(\lambda_k - \lambda_{k_1} - \lambda_{k_2})\right) \Pi_{\lambda_k}\mathcal{B}(\Pi_{\lambda_{k_1}}\tilde{V}_0^N, \Pi_{\lambda_{k_2}}\tilde{V}_0^N)$$
$$+ \varepsilon\mathcal{W}\left(\frac{t}{\varepsilon}\right)\mathcal{B}\left(\mathcal{W}\left(-\frac{t}{\varepsilon}\right)\tilde{V}_1^N, \mathcal{W}\left(-\frac{t}{\varepsilon}\right)(2\tilde{V}_0^N + \varepsilon\tilde{V}_1^N)\right)$$
$$- 2\varepsilon J_N \sum_{\lambda_{k_1} + \lambda_{k_2} \neq \lambda_k} \frac{\exp\left(-\frac{it}{\varepsilon}(\lambda_{k_1} + \lambda_{k_2} - \lambda_k)\right)}{i(\lambda_k - \lambda_{k_1} - \lambda_{k_2})} \Pi_{\lambda_k}\mathcal{B}(\Pi_{\lambda_{k_1}}\partial_t\tilde{V}_0^N, \Pi_{\lambda_{k_2}}\tilde{V}_0^N)$$

$$(5.58)$$

From the uniform bound on $\tilde{V}_0^N$ and the convergence $\tilde{V}_0^N \to \tilde{V}_0$ we deduce that the first three terms in the right-hand side of the previous identity go to zero as $N \to \infty$. For fixed N, using the (non-uniform) estimates on $\tilde{V}_0^N$ and $\tilde{V}_1^N$ and the bound from below on $|\lambda_k - \lambda_{k_1} - \lambda_{k_2}|$, we obtain that the three other terms go to zero as $\varepsilon \to 0$. We conclude that

$$\partial_t(\tilde{V}_0^N + \varepsilon\tilde{V}_1^N) + \mathcal{W}\left(\frac{t}{\varepsilon}\right)\mathcal{B}\left(\mathcal{W}\left(-\frac{t}{\varepsilon}\right)(\tilde{V}_0^N + \varepsilon\tilde{V}_1^N), \mathcal{W}\left(-\frac{t}{\varepsilon}\right)(\tilde{V}_0^N + \varepsilon\tilde{V}_1^N)\right)$$

goes to zero in $L^\infty_{loc}([0, t^*), H^{s-1}(\Omega))$ as $\varepsilon \to 0$ then $N \to \infty$.

In other words $(\tilde{V}_0^N + \varepsilon\tilde{V}_1^N)$ is an approximate solution to (5.52) in the sense of (5.50). However it does not satisfy the (approximate) global conservation of mass (5.51).

Indeed we have proved that the oscillating modes have zero total mass (see (5.57)) and that the equation governing the non-oscillating part of $\tilde{V}_0$ is conservative, from which we deduce that the first three terms in the right-hand side of (5.58) have no contribution to the variation of the total mass. But the last two terms are expected to give some variation of the total mass of order ε, which is not admissible for (5.51) to hold.

• We therefore have to build some third order approximation $\tilde{V}_2^N$, which is done by the same process, using the fact that the only contribution to the total mass comes from the non-oscillating part. We skip that part of the construction and refer to [95] for the details.

Defining

$$(\rho_\varepsilon^N, u_\varepsilon^N, \theta_\varepsilon^N) = \mathcal{W}\left(-\frac{t}{\varepsilon}\right)(\tilde{V}_0^N + \varepsilon\tilde{V}_1^N + \varepsilon^2\tilde{V}_2^N)$$

we then check easily that the uniform bound (5.48) and the convergences (5.49)(5.50) and (5.51) are satisfied.   □

5.4.2 Control of the Flux Term and Proof of Convergence

The conclusion is then very similar to the "well-prepared" case. We start by estimating the momentum and energy fluxes in terms of the modulated entropy and entropy dissipation :

Proposition 5.4.2 *Under the same assumptions as in Theorem 5.1.11, for any $(\tilde{\rho}, \tilde{u}, \tilde{\theta}) \in C_c^\infty([0, t^*] \times \bar{\Omega})$ such that $n \cdot \tilde{u}_{|\partial\Omega} = 0$,*

$$-\frac{1}{\varepsilon^2}\int_0^t \iint f_\varepsilon \nabla_x\tilde{u} : \Phi_\varepsilon dxdvds - \frac{1}{\varepsilon^2}\int_0^t \iint f_\varepsilon e^{\frac{1}{2}\varepsilon\tilde{\theta}}\nabla_x\tilde{\theta} \cdot \Psi_\varepsilon dxdvds$$
$$\leq \frac{C}{\varepsilon^2}\int_0^t \|D_x(\tilde{u}, \tilde{\theta})(s)\|_{L^2 \cap L^\infty(\Omega)} H(f_\varepsilon|\mathcal{M}_\varepsilon)(s)ds + o(1) \qquad (5.59)$$

where the constant C depends only on the L^∞ norm of $(\tilde{\rho}, \tilde{u}, \tilde{\theta})$.

Proof. The proof relies on the same type of decomposition of the fluxes :

$$\frac{1}{\varepsilon^2}\int \Phi_\varepsilon(f_\varepsilon - \mathcal{M}_\varepsilon)(t,x,v)dv = -2\varepsilon^{(q-1)/2}\int \tilde\Phi_\varepsilon\mathcal{M}_\varepsilon\tilde q_\varepsilon(t,x,v)dv$$
$$+\frac{1}{2}\int \tilde\Phi_\varepsilon Q(\mathcal{M}_\varepsilon\tilde g_\varepsilon, \mathcal{M}_\varepsilon\tilde g_\varepsilon)(t,x,v)dv$$
$$+\frac{1}{4}\int \tilde\Phi_\varepsilon\mathcal{M}_\varepsilon\tilde g_\varepsilon^2(t,x,v)dv\,,$$

and

$$\frac{1}{\varepsilon^2}\int \Psi_\varepsilon(f_\varepsilon - \mathcal{M}_\varepsilon)(t,x,v)dv = -2\varepsilon^{(q-1)/2}\int \tilde\Psi_\varepsilon\mathcal{M}_\varepsilon\tilde q_\varepsilon(t,x,v)dv$$
$$+\frac{1}{2}\int \tilde\Psi_\varepsilon Q(\mathcal{M}_\varepsilon\tilde g_\varepsilon, \mathcal{M}_\varepsilon\tilde g_\varepsilon)(t,x,v)dv$$
$$+\frac{1}{4}\int \tilde\Psi_\varepsilon\mathcal{M}_\varepsilon\tilde g_\varepsilon^2(t,x,v)dv\,.$$

The difficulty comes from the fact that $\Psi_\varepsilon = O(|v|^3)$ as $|v| \to \infty$, and that moments of order 3 cannot be controlled by the modulated entropy via Young's inequality.

Instead of the controls (5.41) on large velocities, we will therefore use the additional (non uniform) a priori estimate (5.19) on large tails. We have indeed for all $p < 1$

$$\int \mathcal{M}_\varepsilon\left(\frac{f_\varepsilon}{\mathcal{M}_\varepsilon}\right)^{2p} dv \le \left(\int \frac{f_\varepsilon^2}{M}dv\right)^p \left(\int \left(\frac{M^p}{\mathcal{M}_\varepsilon^{2p-1}}\right)^{1/(1-p)} dv\right)^{1-p}$$
$$\le C_p$$

provided that ε is sufficiently small (depending on p), since the moments of $\mathcal{M}_\varepsilon$ differ from those of M by quantities of order ε. We then have

$$\varepsilon\tilde g_\varepsilon = O(1)_{L^\infty_{t,x}(L^{4p}(\mathcal{M}_\varepsilon dv))} \text{ and thus } \varepsilon(\tilde g_\varepsilon - \Pi\tilde g_\varepsilon) = O(1)_{L^\infty_{t,x}(L^{4p}(\mathcal{M}_\varepsilon dv))}.$$

On the other hand, the relaxation estimate states

$$\tilde g_\varepsilon - \Pi_\varepsilon\tilde g_\varepsilon = \varepsilon^{\frac{q+1}{2}}O\left(\frac{1}{\varepsilon^{\frac{q+3}{2}}}D(f_\varepsilon)^{1/2}\right)_{L^2(\mathcal{M}_\varepsilon dv)} + \varepsilon O\left(\frac{1}{\varepsilon^2}H(f_\varepsilon|\mathcal{M}_\varepsilon)\right)_{L^1_x(L^2(\mathcal{M}_\varepsilon dv))}$$

Coupling both estimates leads finally to the following control on large velocities

$$\tilde g_\varepsilon^2 = \varepsilon^{(q-1)/2}O\left(\frac{1}{\varepsilon^{(q+3)/2}}D(f_\varepsilon)^{1/2}\right)_{L^{\frac{4p}{2p+1}}(\mathcal{M}_\varepsilon dv)}$$
$$+O\left(\frac{1}{\varepsilon^2}H(f_\varepsilon|\mathcal{M}_\varepsilon)\right)_{L^1(dx, L^{\frac{4p}{2p+1}}(\mathcal{M}_\varepsilon dv))} \tag{5.60}$$

From that weighted L^2 estimate on $\tilde g_\varepsilon$ and the bound on $\tilde q_\varepsilon$ coming from the entropy dissipation, we finally deduce that

$$\frac{1}{\varepsilon^2}\int \Phi_\varepsilon(f_\varepsilon - \mathcal{M}_\varepsilon)(t,x,v)dv = \varepsilon^{(q-1)/2}O\left(\frac{1}{\varepsilon^{(q+3)/2}}\left(\int D(f_\varepsilon)dx\right)^{1/2}\right)_{L^2(dx)}$$
$$+O\left(\frac{1}{\varepsilon^2}H(f_\varepsilon|\mathcal{M}_\varepsilon)\right)_{L^1(dx)},$$

and

$$\frac{1}{\varepsilon^2}\int \Psi_\varepsilon(f_\varepsilon - \mathcal{M}_\varepsilon)(t,x,v)dv = \varepsilon^{(q-1)/2}O\left(\frac{1}{\varepsilon^{(q+3)/2}}\left(\int D(f_\varepsilon)dx\right)^{1/2}\right)_{L^2(dx)}$$
$$+O\left(\frac{1}{\varepsilon^2}H(f_\varepsilon|\mathcal{M}_\varepsilon)\right)_{L^1(dx)}.$$

which is the convenient form to apply Gronwall's lemma.

Note that the condition (5.19) is actually a very strong assumption, which could be relaxed since we only need to control the third moment of the distribution. □

We then establish the convergence result stated in Theorem 5.1.11 in the case of well-prepared velocity profiles

$$g_{in} = \rho_{in} + u_{in}\cdot v + \frac{1}{2}\theta_{in}(|v|^2 - 3) \tag{5.61}$$

by the same arguments as in paragraph 5.3.2.

• Combining (5.31) and (5.59), and integrating the resulting differential inequality, we get by Gronwall's lemma

$$\frac{1}{\varepsilon^2}H(f_\varepsilon|\mathcal{M}_\varepsilon)(t) + \frac{1}{\varepsilon^{q+3}}\int_0^t\int D(f_\varepsilon)dxds$$
$$\leq \frac{1}{\varepsilon^2}H(f_{\varepsilon,in}|\mathcal{M}_{\varepsilon,in})\exp\left(C\int_0^t\left(\|\nabla_x\tilde{u}\|_{L^2\cap L^\infty}+\|\nabla_x\tilde{\theta}\|_{L^2\cap L^\infty}\right)ds\right)+o(1)$$
$$-\int_0^t\exp\left(C\int_s^t\left(\|\nabla_x\tilde{u}\|_{L^2\cap L^\infty}+\|\nabla_x\tilde{\theta}\|_{L^2\cap L^\infty}\right)d\tau\right)$$
$$\iint\left(1,e^{-\varepsilon\tilde{\theta}}(v-\varepsilon\tilde{u}),\frac{1}{2}\left(e^{-\varepsilon\tilde{\theta}}|v-\varepsilon\tilde{u}|^2-3\right)\right)\cdot\mathbf{A}_\varepsilon(\tilde{\rho},\tilde{u},\tilde{\theta})f_\varepsilon dvdxds \tag{5.62}$$

for any $(\tilde{\rho},\tilde{u},\tilde{\theta}) \in C_c^\infty(\mathbf{R}^+\times\Omega)$ such that $n\cdot\tilde{u}_{|\partial\Omega} = 0$.

• Plugging the approximate solution $(\rho_\varepsilon^N,u_\varepsilon^N,\theta_\varepsilon^N)$ built in Proposition 5.4.1 in Gronwall's inequality (5.62)

$$\mathcal{M}_\varepsilon^N = \mathcal{M}_{(\exp(\varepsilon\rho_\varepsilon^N),\varepsilon u_\varepsilon^N,\exp(\varepsilon\theta_\varepsilon^N))}$$

leads then to

$$\frac{1}{\varepsilon^2} H(f_\varepsilon | \mathcal{M}_\varepsilon^N) \to 0 \text{ in } L_{loc}^\infty([0, t^*)) \text{ as } \varepsilon \to 0 \text{ then } N \to \infty. \tag{5.63}$$

We first prove that the first term in the right-hand side of (5.62) converges to 0 as $\varepsilon \to 0$ then $N \to \infty$. Denoting

$$\mathcal{M}_{\varepsilon, in} = \mathcal{M}_{(\exp(\varepsilon \rho_{in}), \varepsilon u_{in}, \exp(\varepsilon \theta_{in}))}$$

we have the identity

$$H(f_{\varepsilon, in} | \mathcal{M}_{\varepsilon, in}^N) = H\left(f_{\varepsilon, in} | \mathcal{M}_{\varepsilon, in}\right) + H\left(\mathcal{M}_{\varepsilon, in} | \mathcal{M}_{\varepsilon, in}^N\right)$$
$$+ \iint (f_{\varepsilon, in} - \mathcal{M}_{\varepsilon, in}) \log \frac{\mathcal{M}_{\varepsilon, in}}{\mathcal{M}_{\varepsilon, in}^N} dv dx$$

We therefore obtain (5.63) from the assumptions on the initial data (5.18) (5.61) and the convergence statement (5.49)

$$(\rho_{\varepsilon, in}^N, u_{\varepsilon, in}^N, \theta_{\varepsilon, in}^N) \to (\rho_{in}, u_{in}, \theta_{in})$$

which implies in particular that

$$\frac{1}{\varepsilon^2} H\left(\mathcal{M}_{\varepsilon, in} | \mathcal{M}_{\varepsilon, in}^N\right) \to 0.$$

The convergence of the two other terms in the right-hand side of (5.62) is obtained by combining the uniform bound (with respect to ε and N)

$$\|D_x(u_\varepsilon^N, \theta_\varepsilon^N)\|_{L^\infty([0,t], L^2 \cap L^\infty(\Omega))} \le C$$

with the convergence statements (5.50) and (5.51) :

$$\frac{1}{\varepsilon^2} \int \partial_t \exp(\varepsilon \rho_\varepsilon^N) dx \to 0, \quad A_\varepsilon(\rho_\varepsilon^N, u_\varepsilon^N, \theta_\varepsilon^N) \to 0.$$

• From the entropic convergence (5.63) we deduce the convergence of the moments stated in Theorem 5.1.11 by the same type of functional inequalities as in paragraph 5.3.2.

We indeed recall that, by Lemma 3.1.2 in Chapter 3, up to extraction of a subsequence, we have the convergences

$$\rho_\varepsilon = \frac{1}{\varepsilon} \int (f_\varepsilon - M) dv \to \rho \text{ weakly in } L_{loc}^\infty(dt, L_{loc}^1(dx)),$$

$$u_\varepsilon = \frac{1}{\varepsilon} \int f_\varepsilon v dv \rightharpoonup u \text{ weakly in } L_{loc}^\infty(dt, L_{loc}^1(dx)),$$

$$\theta_\varepsilon = \frac{1}{3\varepsilon}\int (f_\varepsilon - M)(|v|^2 - 3)dv \rightharpoonup \theta \text{ weakly in } L^\infty_{loc}(dt, L^1_{loc}(dx)).$$

Furthermore, taking limits in the local conservations of mass and momentum

$$\partial_t \int f_\varepsilon dv + \frac{1}{\varepsilon}\nabla_x \cdot \int f_\varepsilon v dv = 0$$

$$\partial_t \int f_\varepsilon v dv + \frac{1}{\varepsilon}\nabla_x \cdot \int f_\varepsilon v^{\otimes 2} dv = 0$$

shows that

$$\nabla_x \cdot u = 0, \quad \nabla_x(\rho + \theta) = 0.$$

We further have

$$\frac{1}{\varepsilon^2}H(f_\varepsilon|\mathcal{M}_\varepsilon) \geq \frac{1}{\varepsilon^2}H(\mathcal{M}_{f_\varepsilon}|\mathcal{M}_\varepsilon)$$

$$\geq \int \exp(\varepsilon\rho^N_\varepsilon)h\left(R_\varepsilon \exp(-\varepsilon\rho^N_\varepsilon) - 1\right)(t,x)dx$$

$$+ \frac{1}{2}\int R_\varepsilon e^{-\varepsilon\theta^N_\varepsilon}|U_\varepsilon - \varepsilon u^N_\varepsilon|^2(t,x)dx$$

$$+ \frac{3}{2}\int R_\varepsilon\left(T_\varepsilon \exp(-\varepsilon\theta^N_\varepsilon) - \log(T_\varepsilon \exp(-\varepsilon\theta^N_\varepsilon)) - 1\right)(t,x)dx$$

with

$$R_\varepsilon = 1 + \varepsilon\rho_\varepsilon, \quad U_\varepsilon = \varepsilon u_\varepsilon + O(\varepsilon^2) \text{ and } T_\varepsilon = 1 + \varepsilon\theta_\varepsilon + O(\varepsilon^2).$$

We therefore have

$$\rho_\varepsilon - \rho^N_\varepsilon \to 0 \text{ strongly in } L^\infty_{loc}(dt, L^1_{loc}(dx)),$$

$$u_\varepsilon - u^N_\varepsilon \to 0 \text{ strongly in } L^\infty_{loc}(dt, L^1_{loc}(dx)),$$

$$\theta_\varepsilon - \theta^N_\varepsilon \to 0 \text{ strongly in } L^\infty_{loc}(dt, L^1_{loc}(dx)).$$

Moreover, by construction of $(\rho^N_\varepsilon, u^N_\varepsilon, \theta^N_\varepsilon)$, we know that it converges weakly to the solution $(\bar\rho, \bar u, \bar\theta)$ of the incompressible Euler equations (5.56).

This concludes the proof in the case when the initial data is close to thermodynamic equilibrium (5.61), i.e. in the case when there is no relaxation layer.

5.5 Taking into Account the Knudsen Layer

For general initial data, the purely kinetic part of the solution to the Boltzmann equation is expected to converge to 0 exponentially in time, in particular in $L^1_{loc}(dtdxdv)$, but not in $L^\infty_{loc}(dt, L^1_{loc}(dxdv))$. In order to consider the relaxation process in the relative entropy method, one thus has to construct a refined approximation f_{app}, and then to introduce it in the modulated entropy inequality (5.31). This requires in particular to also modulate the entropy dissipation.

5.5.1 The Refined Stability Inequality

Proposition 5.5.1 *Denote by f_{app} some smooth function on $[0, t^*] \times \Omega \times \mathbf{R}^3$ satisfying the boundary condition*

$$f_{app}(t, x, v) = f_{app}(t, x, v - 2(v \cdot n(x))n(x)) \text{ on } \Sigma^-.$$

Then, any solution to the scaled Boltzmann equation (5.17) such that (5.19) holds satisfies the following modulated entropy inequality

$$H(f_\varepsilon \,|f_{app})(t) + \frac{1}{\varepsilon^{q+1}} \int_0^t \int D(f_\varepsilon |f_{app}) ds dx \leq H(f_\varepsilon^{in} |f_{app}^{in})$$
$$- \varepsilon \int_0^t \iint \tilde{g}_\varepsilon \left(\partial_t f_{app} + \frac{1}{\varepsilon} v \cdot \nabla_x f_{app} - \frac{1}{\varepsilon^{q+1}} Q(f_{app}, f_{app}) \right) dv dx ds$$
$$+ \frac{1}{4\varepsilon^{q-1}} \int_0^t \iiint (\tilde{g}_\varepsilon \tilde{g}_{\varepsilon 1} - \tilde{g}'_\varepsilon \tilde{g}'_{\varepsilon 1})(f'_{app} f'_{app1} - f_{app} f_{app1}) B dv dv_1 d\sigma dx ds$$

(5.64)

where $\tilde{g}_\varepsilon$ denotes the fluctuation

$$\tilde{g}_\varepsilon = \frac{1}{\varepsilon} \frac{f_\varepsilon - f_{app}}{f_{app}},$$

and $D(f_\varepsilon | f_{app})$ is the modulated entropy dissipation defined by

$$D(f_\varepsilon | f_{app}) = \frac{1}{4} \iint (f'_\varepsilon f'_{\varepsilon 1} - f_\varepsilon f_{\varepsilon 1}) \log \left(\frac{f'_\varepsilon f'_{\varepsilon 1} f_{app} f_{app1}}{f_\varepsilon f_{\varepsilon 1} f'_{app} f'_{app1}} \right)$$
$$- (f'_{app} f'_{app1} - f_{app} f_{app1}) \left(\frac{f'_\varepsilon f'_{\varepsilon 1}}{f'_{app} f'_{app1}} - \frac{f_\varepsilon f_{\varepsilon 1}}{f_{app} f_{app1}} \right) B dv dv_1 d\sigma.$$

(5.65)

Remark 5.5.2 *Note that the integrand arising in the definition of the modulated entropy dissipation is always nonnegative, which is crucial to get some stability. We have indeed*

$$D(f_\varepsilon | f_{app}) = \frac{1}{4} \iiint f_\varepsilon f_{\varepsilon 1} \left(k \left(\frac{f'_\varepsilon f'_{\varepsilon 1}}{f_\varepsilon f_{\varepsilon 1}} \right) - k \left(\frac{f'_{app} f'_{app1}}{f_{app} f_{app1}} \right) \right.$$
$$\left. - \left(\frac{f'_\varepsilon f'_{\varepsilon 1}}{f_\varepsilon f_{\varepsilon 1}} - \frac{f'_{app} f'_{app1}}{f_{app} f_{app1}} \right) k' \left(\frac{f'_{app} f'_{app1}}{f_{app} f_{app1}} \right) \right) B dv dv_1 d\sigma$$

where k is the convex function defined by $k(z) = (z - 1) \log z$.

This has naturally to be compared with the definition (5.14) of the modulated entropy

$$H(f_\varepsilon | f_{app}) = \iint f_{app} \left(h(f_\varepsilon - 1) - h(f_{app} - 1) - (f_\varepsilon - f_{app}) h'(f_{app} - 1) \right) dv dx$$

with $h(z) = (1 + z) \log(1 + z) - z$.

Proof. Start from the entropy inequality (5.32) satisfied by the solution of the scaled Boltzmann equation with specular reflection at the boundary:

$$H(f_\varepsilon(t)|M) + \frac{1}{\varepsilon^{q+1}} \int_0^t \int D(f_\varepsilon)(s,x)dsdx \le H(f_{\varepsilon,in}|M)$$

By definition of the modulated entropy (5.14), we then have

$$H\,(f_\varepsilon|f_{app})(t) + \frac{1}{\varepsilon^{q+1}} \int_0^t \int D(f_\varepsilon)dsdx$$

$$\le H(f_{\varepsilon,in}|f_{app,in}) - \int_0^t \iint \log f_{app}\left(\partial_t + \frac{1}{\varepsilon}v\cdot\nabla_x\right)f_\varepsilon dvdxds \qquad (5.66)$$

$$- \int_0^t \iint \left(\frac{f_\varepsilon}{f_{app}} - 1\right)\left(\partial_t + \frac{1}{\varepsilon}v\cdot\nabla_x\right)f_{app}dvdxds$$

using the specular reflection on the boundary for f_ε and f_{app}.

Now, for solutions of the Boltzmann equation satisfying (5.19), the collision term $Q(f_\varepsilon, f_\varepsilon)$ makes sense, and the kinetic equation (5.17) holds in the sense of distributions. We thus have

$$-\int_0^t \iint \log f_{app}\left(\partial_t + \frac{1}{\varepsilon}v\cdot\nabla_x\right)f_\varepsilon dvdxds$$

$$= -\frac{1}{\varepsilon^{q+1}}\int_0^t \iint (\log f_{app})Q(f_\varepsilon, f_\varepsilon)dvdxds$$

$$= -\frac{1}{4\varepsilon^{q+1}}\int_0^t \iint \iint (f'_\varepsilon f'_{\varepsilon 1} - f_\varepsilon f_{\varepsilon 1})\log\left(\frac{f_{app}f_{app1}}{f'_{app}f'_{app1}}\right)Bdvdv_1d\sigma dxds$$

using the classical symmetries of the collision integrand.

In the same way, we have

$$-\int_0^t \iint \left(\frac{f_\varepsilon}{f_{app}} - 1\right)\left(\partial_t + \frac{1}{\varepsilon}v\cdot\nabla_x\right)f_{app}dvdxds$$

$$= -\int_0^t \iint \frac{f_\varepsilon - f_{app}}{f_{app}}\left(\partial_t f_{app} + \frac{1}{\varepsilon}v\cdot\nabla_x f_{app} - \frac{1}{\varepsilon^{q+1}}Q(f_{app}, f_{app})\right)dvdxds$$

$$+ \frac{1}{4\varepsilon^{q+1}}\int_0^t \iiint (f'_{app}f'_{app1} - f_{app}f_{app1})$$

$$\left(\frac{f'_\varepsilon}{f'_{app}} + \frac{f'_{\varepsilon 1}}{f'_{app1}} - \frac{f_\varepsilon}{f_{app}} - \frac{f_{\varepsilon 1}}{f_{app1}}\right)Bdvdv_1d\sigma dxds$$

$$= -\int_0^t \iint \frac{f_\varepsilon - f_{app}}{f_{app}}\left(\partial_t f_{app} + \frac{1}{\varepsilon}v\cdot\nabla_x f_{app} - \frac{1}{\varepsilon^{q+1}}Q(f_{app}, f_{app})\right)dvdxds$$

$$+ \frac{1}{4\varepsilon^q}\int_0^t \iiint (f'_{app}f'_{app1} - f_{app}f_{app1})(\tilde{g}'_\varepsilon + \tilde{g}'_{\varepsilon 1} - \tilde{g}_\varepsilon - \tilde{g}_{\varepsilon 1})Bdvdv_1d\sigma dxds$$

Plugging both identities in (5.66) leads then to

$$
H\left(f_\varepsilon|f_{app}\right)(t) + \frac{1}{\varepsilon^{q+1}} \int_0^t \int D(f_\varepsilon|f_{app})\,ds\,dx \le H(f_{\varepsilon,in}|f_{app,in})
$$
$$
- \int_0^t \iint \frac{f_\varepsilon - f_{app}}{f_{app}} \left(\partial_t f_{app} + \frac{1}{\varepsilon} v \cdot \nabla_x f_{app} - \frac{1}{\varepsilon^{q+1}} Q(f_{app}, f_{app}) \right) dv\,dx\,ds
$$
$$
+ \frac{1}{4\varepsilon^{q+1}} \int_0^t \iiiint (f'_{app} f'_{app1} - f_{app} f_{app1})(\tilde{g}_\varepsilon \tilde{g}_{\varepsilon 1} - \tilde{g}'_\varepsilon \tilde{g}'_{\varepsilon 1}) B\,dv\,dv_1\,d\sigma\,dx\,ds
$$

which is the expected inequality.   □

5.5.2 Construction of An Approximate Solution

The next step is to construct suitable approximate solutions f_{app}. Let us recall that, in the initial layer, the dominating process is expected to be the relaxation, so that the transport can be neglected in first approximation.

We thus solve the homogeneous equation

$$
\partial_t f = \frac{1}{\varepsilon^{q+1}} Q(f, f),
$$
$$
f(x)_{|t=0} = f_{in}(x).
$$
(5.67)

using a fixed point argument in some functional space with exponential time decay (see [59] for the well-posedness of (5.67) and Appendix B2 of [95] for the suitable relaxation estimate).

One of the difficulty here is that we further need to control the dependence with respect to the spatial variable x, to get some uniform bound on

$$
\frac{1}{\varepsilon^2} \int_0^{\tau_\varepsilon} \|v \cdot \nabla_x \log f_{app}(s)\|_{L^\infty(\Omega, L^{p'}(f_{app}dv))}\,ds
$$

for some τ_ε characterizing the size of the initial layer and some $p' > 2$ to be determined later.

What can be proved is the following

Proposition 5.5.3 *Let $(f_{\varepsilon,in})$ be some sequence of initial data such that the sequence of fluctuations $(g_{\varepsilon,in})$ defined by $f_{\varepsilon,in} = M(1 + \varepsilon g_{\varepsilon,in})$ converges entropically to some $g_{in} \in L^2(Mdvdx)$, i.e. such that*

$$
\frac{1}{\varepsilon^2} H(f_{\varepsilon,in}|M) \to \iint M(g_{in})^2 dx dv.
$$
(5.68)

Then there exists some family (f_ε^N) of nonnegative functions satisfying approximatively the homogeneous Boltzmann equation as $\varepsilon \to 0$ then $N \to \infty$

$$
\partial_t f_\varepsilon^N - \frac{1}{\varepsilon^{q+1}} Q(f_\varepsilon^N, f_\varepsilon^N) \to 0 \text{ in } L^2(dtdx, L^{p'}(f_\varepsilon^N dv))
$$
(5.69)

with suitable initial data

$$\frac{1}{\varepsilon^2} H(f_{\varepsilon,in}|f_{\varepsilon,in}^N) \to 0. \qquad (5.70)$$

It furthermore satisfies the relaxation estimate

$$\frac{1}{\varepsilon^{q+1}} \int_0^t \left\| \frac{f_\varepsilon^{N\prime} f_{\varepsilon,1}^{N\prime}}{f_\varepsilon^N f_{\varepsilon,1}^N} - 1 \right\|_{L^\infty(\Omega, L^{p\prime}(f_\varepsilon^N f_{\varepsilon,1}^N B dv dv_1 d\sigma))} ds \to 0, \qquad (5.71)$$

and the regularity estimate

$$\left\| v \cdot \nabla_x \log f_\varepsilon^N \right\|_{L^2(dx, L^{p\prime}(f_\varepsilon^N dv))} \le C\varepsilon \ \textit{uniformly in time}, \qquad (5.72)$$

for some $p\prime > 4$.

Sketch of the Proof of Proposition 5.5.3. The proof of the previous result is very technical and can be found in [95]. We just give here the main ideas.

In order to build the approximate solution we need, we start from the solution of the homogeneous Boltzmann equation (5.67) with some smooth initial data, then truncate the contribution of large velocities to get a Maxwellian bound from below on f_ε^N.

• The first step consists therefore in choosing some suitable initial data. In order that the solution to (5.67) has good decay properties with respect to v and a smooth dependence with respect to x (which is now a simple parameter), it is enough to impose such conditions on the initial data.

The approximate initial data $f_{\varepsilon,in}^N$ is therefore obtained by some spatial regularization and some truncation of large velocities, such that

$$\frac{1}{\varepsilon^2} H(f_{\varepsilon,in}|f_{\varepsilon,in}^N) \to 0 \ \text{as} \ \varepsilon \to 0 \ \text{then} \ N \to \infty.$$

Standard results on the collision operator ensure then that the regularity in x, and decay in v are propagated, and that the corresponding solution $\tilde{f}_\varepsilon^N$ of (5.67) tends exponentially in (t/ε^{q+1}) to local equilibrium.

• In order to have a bound from below on f_ε^N, we then need to truncate large velocities

$$f_\varepsilon^N = (\tilde{f}_\varepsilon^N - \mathcal{M}_{\tilde{f}_\varepsilon^N}) 1_{|v|^2 \le K|\log \varepsilon|} + \mathcal{M}_{\tilde{f}_\varepsilon^N}.$$

For ε sufficiently small, we can prove that the approximate solution f_ε^N satisfies

$$f_{\varepsilon,in}^N = \tilde{f}_{\varepsilon,in}^N$$

and satisfies both (5.71) and (5.72).

However the homogeneous Boltzmann equation (5.67) is no longer satisfied : it remains then to check that (5.69) holds. The conclusion follows from standard estimates on the tails of Maxwellian distributions and a suitable choice of the truncation parameter K.   □

5.5.3 Control of the Flux Term and Proof of Convergence

The previous estimates on the sequence of approximate solutions to the homogeneous Boltzmann equation (5.67) should allow to control the different terms in the entropy inequality (5.64) provided that the fluctuation $\tilde{g}_\varepsilon$ defined by

$$\tilde{g}_\varepsilon = \frac{1}{\varepsilon} \frac{f_\varepsilon - f_\varepsilon^N}{f_\varepsilon^N},$$

can be estimated in terms of the scaled relative entropy $\varepsilon^{-2} H(f_\varepsilon | f_\varepsilon^N)$ in a suitable norm.

• Because f_ε^N is bounded from up and below by some Maxwellians, the same arguments as in the proof of Proposition 5.4.2 show that, under the assumption (5.19), for all $p < \frac{4}{3}$, there exists $C_p > 0$ such that

$$\|\tilde{g}_\varepsilon\|^2_{L^2(\Omega, L^p(f_\varepsilon^N dv))} \leq \frac{C_p}{\varepsilon^2} H(f_\varepsilon | f_\varepsilon^N) + o(1). \tag{5.73}$$

Equipped with these preliminary results, we are now able to achieve the proof of Theorem 5.1.11 in the general case. Actually we will prove that on a thin time layer, the distribution becomes close to local thermodynamic equilibrium in the following sense

$$\frac{1}{\varepsilon^2} H(f_\varepsilon(\tau_\varepsilon) | \mathcal{M}_{\varepsilon, in}) \to 0 \text{ for some } \tau_\varepsilon \to 0 \text{ as } \varepsilon \to 0,$$

where $\mathcal{M}_{\varepsilon, in}$ is the initial local thermodynamic equilibrium

$$\mathcal{M}_{\varepsilon, in} = \mathcal{M}_{(\exp(\varepsilon \rho_{in}), \varepsilon u_{in}, \exp(\varepsilon \theta_{in}))}.$$

Then, we will use the results of the previous section, combined with the continuity with respect to time of the solutions to the system (5.45), to obtain the convergence on the time interval $[\tau_\varepsilon, t^*)$.

• Combining Proposition 5.5.1 and estimate (5.73) leads to the following inequality

$$\frac{1}{\varepsilon^2} H(f_\varepsilon | f_\varepsilon^N)(t) + \frac{1}{\varepsilon^{q+3}} \int_0^t \int D(f_\varepsilon | f_\varepsilon^N) ds dx \leq \frac{1}{\varepsilon^2} H(f_{\varepsilon, in} | f_{\varepsilon, in}^N)$$

$$- \frac{1}{\varepsilon} \int_0^t \iint \tilde{g}_\varepsilon \left(\partial_t f_\varepsilon^N + \frac{1}{\varepsilon} v \cdot \nabla_x f_\varepsilon^N - \frac{1}{\varepsilon^{q+1}} Q(f_\varepsilon^N, f_\varepsilon^N) \right) dv dx ds$$

$$+ \frac{C_p}{2\varepsilon^{q+1}} \int_0^t \left\| \frac{f_\varepsilon^{N'} f_{\varepsilon,1}^{N'}}{f_\varepsilon^{N'} f_{\varepsilon,1}^N} - 1 \right\|_{L^\infty(\Omega, L^{p'}(f_\varepsilon^N f_{\varepsilon,1}^N, B dv dv_1 d\sigma))} \frac{1}{\varepsilon^2} H(f_\varepsilon | f_\varepsilon^N)(s) ds$$

Integrating next this differential inequality leads to

$$\frac{1}{\varepsilon^2}H(f_\varepsilon|f_\varepsilon^N)(t) + \frac{1}{\varepsilon^{q+3}}\int_0^t\int\int D(f_\varepsilon|f_\varepsilon^N)dvdxds$$

$$\leq \frac{1}{\varepsilon^2}H(f_{\varepsilon,in}|f_{\varepsilon,in}^N)\exp\left(\chi_\varepsilon^N(0,t)\right)$$

$$+\frac{1}{\varepsilon}\int_0^t\exp\left(\chi_\varepsilon^N(s,t)\right)\int\int\tilde{g}_\varepsilon\left(\partial_t f_\varepsilon^N + \frac{1}{\varepsilon}v\cdot\nabla_x f_\varepsilon^N - \frac{1}{\varepsilon^{q+1}}Q(f_\varepsilon^N,f_\varepsilon^N)\right)dvdxds$$

$$(5.74)$$

where χ_ε^N is the function defined by

$$\chi_\varepsilon^N(s,t) = \frac{C_p}{2\varepsilon^{q+1}}\int_s^t\left\|\left|\frac{f_\varepsilon^{N'}f_{\varepsilon,1}^{N'}}{f_\varepsilon^N f_{\varepsilon,1}^N} - 1\right|\right\|_{L^\infty(\Omega,L^{p'}(f_\varepsilon^N f_{\varepsilon,1}^N Bdvdv_1 d\sigma))}ds.$$

Using the estimates on f_ε^N established in Proposition 5.5.3 leads then to

$$\sup_{t\in[0,\tau_\varepsilon]}\frac{1}{\varepsilon^2}H(f_\varepsilon|f_\varepsilon^N)(t) \to 0 \qquad (5.75)$$

for any τ_ε such that

$$\frac{\varepsilon^{q+1}}{\tau_\varepsilon} \to 0 \text{ and } \frac{\tau_\varepsilon}{\varepsilon} \to 0.$$

Indeed the first term in the right-hand side of (5.74) is proved to converge to 0 as $\varepsilon \to 0$ using the convergence of the initial data (5.18) and the uniform bound (5.71) :

$$\frac{1}{\varepsilon^2}H(f_{\varepsilon,in}|f_{\varepsilon,in}^N)\exp\left(\chi_\varepsilon(0,t)\right) \to 0.$$

The convergence of the other term in the right-hand side of (5.74) is obtained by combining the uniform bound (5.71) with the convergence statement (5.69) and the regularity estimate (5.72)

$$\frac{1}{\varepsilon}\int_0^t\exp\left(\chi_\varepsilon^N(s,t)\right)\int\int\tilde{g}_\varepsilon\left(\partial_t f_\varepsilon^N - \frac{1}{\varepsilon^{q+1}}Q(f_\varepsilon^N,f_\varepsilon^N)\right)dvdxds \to 0,$$

$$\frac{1}{\varepsilon}\int_0^t\exp\left(\chi_\varepsilon^N(s,t)\right)\int\int\tilde{g}_\varepsilon\left(\frac{1}{\varepsilon}v\cdot\nabla_x f_\varepsilon^N\right)dvdxds \to 0.$$

• We choose for instance

$$\tau_\varepsilon = \varepsilon^q$$

as the upper bound for the relaxation layer.

Inside the relaxation layer, we use the previous arguments which prove that f_ε remains close to f_ε^N. In particular, for $t = \tau_\varepsilon$, the distribution is close to thermodynamic equilibrium, meaning that it satisfies (5.61).

Outside from the relaxation layer, we are then brought back to the situation when the velocity profile is well-prepared, situation we have dealt with in Section 4.4. We therefore define $\mathcal{M}_\varepsilon$ on $[\tau_\varepsilon, t^*)$ by

$$\mathcal{M}_\varepsilon^N = \mathcal{M}_{(\exp(\varepsilon\rho_\varepsilon^N),\varepsilon u_\varepsilon^N,\exp(\varepsilon\theta_\varepsilon^N))}$$

with the following continuity condition for the moments

$$(\rho_\varepsilon^N, u_\varepsilon^N, \theta_\varepsilon^N)(\tau_\varepsilon) = \lim_{t\to\tau_\varepsilon^-} (\rho_\varepsilon^N, u_\varepsilon^N, \theta_\varepsilon^N)(t).$$

Defining $(\rho_\varepsilon^N, u_\varepsilon^N, \theta_\varepsilon^N)$ on $[\tau_\varepsilon, t^*)$ as in Proposition 5.4.1, we then have

$$\frac{1}{\varepsilon^2} H(f_\varepsilon|\mathcal{M}_\varepsilon^N)(t) + \frac{1}{\varepsilon^{q+3}} \int_{\tau_\varepsilon}^t \int D(f_\varepsilon) dx ds$$

$$\leq \frac{1}{\varepsilon^2} H(f_\varepsilon|\mathcal{M}_\varepsilon^N)(\tau_\varepsilon) \exp\left(C \int_{\tau_\varepsilon}^t \left(\|D_x(u_\varepsilon^N, \theta_\varepsilon^N)\|_{L^2\cap L^\infty}\right) ds\right) + o(1)$$

$$- \int_{\tau_\varepsilon}^t \exp\left(C \int_s^t \left(\|D_x(u_\varepsilon^N, \theta_\varepsilon^N)\|_{L^2\cap L^\infty}\right) d\tau\right)$$

$$\int\int \left(1, e^{-\varepsilon\theta_\varepsilon^N}(v - \varepsilon u_\varepsilon^N), \frac{1}{2}\left(e^{-\varepsilon\theta_\varepsilon^N} |v - \varepsilon u_\varepsilon^N|^2 - 3\right)\right) \cdot \mathbf{A}_\varepsilon(\rho_\varepsilon^N, u_\varepsilon^N, \theta_\varepsilon^N) f_\varepsilon dv dx ds$$

$$\tag{5.76}$$

from which we deduce that

$$\sup_{s\in[\tau_\varepsilon, t]} \frac{1}{\varepsilon^2} H(f_\varepsilon|\mathcal{M}_\varepsilon^N) \to 0 \text{ for all } t < t^*, \tag{5.77}$$

provided that the first term in the right-hand side converges to 0 as $\varepsilon \to 0$ then $N \to \infty$.

It remains therefore to check that

$$\frac{1}{\varepsilon^2} H(f_\varepsilon|\mathcal{M}_\varepsilon^N)(\tau_\varepsilon) \to 0$$

which results from the estimate obtained in the first step. We indeed have

$$\frac{1}{\varepsilon^2} H(f_\varepsilon|f_\varepsilon^N)(\tau_\varepsilon) \to 0$$

and, by Proposition 5.5.3,

$$\frac{f_\varepsilon^N}{\mathcal{M}_\varepsilon^N}(\tau_\varepsilon) = 1 + O\left(\varepsilon \exp\left(-\frac{\tau_\varepsilon}{\varepsilon^{q+1}}\right)\right)_{L^\infty},$$

from which we deduce that

$$\frac{1}{\varepsilon^2} H(f_\varepsilon|\mathcal{M}_\varepsilon^N)(\tau_\varepsilon) = \frac{1}{\varepsilon^2} H(f_\varepsilon|f_\varepsilon^N)(\tau_\varepsilon) + \frac{1}{\varepsilon^2} \int\int f_\varepsilon \log\frac{f_\varepsilon^N}{\mathcal{M}_\varepsilon^N}(\tau_\varepsilon) dx dv \to 0$$

as $\varepsilon \to 0$ then $N \to \infty$.

• The entropic convergences (5.75)(5.77) imply the strong convergence of the fluctuation $\tilde{g}_\varepsilon$ in $L_{loc}^\infty([0, t^*), L^1(M dx dv))$ as $\varepsilon \to 0$ then $N \to \infty$.

Finally, using the fact that the purely kinetic part of the approximate solution is equal to 0 on $[\tau_\varepsilon, t^*)$ and converges to 0 in $L_{loc}^1([0, t^*), L^1(dx dv))$, we get the following convergences as $\varepsilon \to 0$:

$$g_\varepsilon - g - g_{osc} \to 0 \text{ in } L_{loc}^\infty((0, t^*), L^1(dx dv)) \cap L_{loc}^1([0, t^*), L^1(dx dv)).$$

6

The Compressible Euler Limit

The last chapter of this survey is devoted to the compressible Euler limit, and is actually a series of remarks and open problems more than a compendium of results.

We will discuss some perspectives regarding the mathematical treatment of this asymptotics. Slight adaptations of the modulated entropy method presented in the previous chapter should give likewise the local convergence towards smooth solutions to the compressible Euler equations under some integrability assumption on the solutions to the Boltzmann equation.

We further hope that suitable improvements of the modulated entropy method (including the modulation of the entropy dissipation) should provide the global convergence of weak solutions to the Boltzmann equation towards entropic solutions to the Riemann problem in one space dimension. The main challenge is of course to understand how the entropy dissipation concentrates on shocks and discontinuities.

6.1 Mathematical Theories for the Compressible Euler System

Let us first recall that the compressible Euler equations constitute a *system of conservation laws*

$$
\begin{aligned}
&\partial_t R + \nabla_x \cdot (RU) = 0, \\
&\partial_t (RU) + \nabla_x \cdot (RU \otimes U + RT Id) = 0, \\
&\partial_t (R|U|^2 + 3RT) + \nabla_x \cdot \big(U(R|U|^2 + 5RT)\big) = 0,
\end{aligned}
\tag{6.1}
$$

insofar as the density R, momentum RU and energy $\frac{1}{2}(R|U|^2 + 3RT)$, which characterize the state of the fluid, are conserved, and that their fluxes

L. Saint-Raymond, *Hydrodynamic Limits of the Boltzmann Equation*,
Lecture Notes in Mathematics 1971, DOI: 10.1007/978-3-540-92847-8_6,
© Springer-Verlag Berlin Heidelberg 2009

depend only on the state of the fluid (in particular, microscopic dissipative effects are neglected). Denoting by V the five-components vector field $V = (R, RU, R|U|^2 + 3RT)$, we can therefore recast the system (6.1) in the abstract form

$$\partial_t V + \nabla_x \cdot F(V) = 0 .$$

That system is *hyperbolic* since the flux matrix

$$DF(V) \cdot \omega \text{ is diagonalizable in } \mathbf{R}^3 \text{ for any } \omega \in \mathbf{R}^3 . \qquad (6.2)$$

Because (6.1) admits an entropy, namely $\log(R/T^{3/2})$, the system is further *symmetrizable*, meaning that there exists some positive definite matrix $A_0(V)$ depending smoothly on V such that

$$A_0 DF(V) \cdot \omega \text{ is symmetric for any } \omega \in \mathbf{R}^3 . \qquad (6.3)$$

This property, coming from an observation of Friedrichs [35] is actually much stronger than the hyperbolicity condition (6.2).

The goal of this first part is to give briefly an idea of the mathematical tools used to study such hyperbolic systems. For more details, we refer to text books such as [101], [36] or [97].

6.1.1 Local Smooth Solutions

The results by Friedrichs [35] show actually that all hyperbolic systems are locally well-posed.

Theorem 6.1.1 *Let* $(R_{in}, U_{in}, T_{in}) \in H^s(\mathbf{R}^3)$ *(*$s > \frac{5}{2}$*) be some given initial density, bulk velocity and temperature with* R_{in} *and* T_{in} *bounded from below.*

Then there exists some $t^* > 0$ *and some* $(R, U, T) \in C^0([0, t^*), H^s(\mathbf{R}^3))$ *such that* (R, U, T) *is a strong solution of (6.1) on* $[0, t^*)$*.*

Except for very particular initial data, the maximal time t^* of existence of such a smooth solution is finite (see [100] for instance). In general, the blow-up corresponds to the apparition of a discontinuity (also called *singularity*). Note that, in the case of the Euler equations, since the hyperbolicity of the system is lost for instance if the density vanishes (cavitation), there could be other sources of blow-up.

Remark 6.1.2 *The previous result can be extended without difficulty to the torus* $\mathbf{T}^3$*, or even to domains with boundaries provided that the boundary conditions are smooth and satisfy some compatibility conditions (only incoming fluxes can be prescribed).*

6.1.2 Weak and Entropic Solutions

If one allows solutions to be discontinuous (considering for instance fields in L^∞ satisfying (6.1) in the sense of distributions), then there is no more uniqueness, nor stability of solutions with respect to initial data. More precisely, one can find sequences of solutions to (6.1) converging in $w * -L^\infty$ towards functions which do not satisfy the equations.

In order to retrieve the stability of solutions, one has therefore to impose additional conditions on the weak solutions.

Definition 6.1.3 *An entropic solution to the compressible Euler equations (6.1) is a field*

$$(R, U, T) \in L^\infty([0, t^*) \times \mathbb{R}^3)$$

satisfying (6.1) in the sense of distributions, as well as the entropy inequality

$$\partial_t \left(R \log \frac{R}{T^{3/2}} \right) + \nabla_x \cdot \left(RU \log \frac{R}{T^{3/2}} \right) \leq 0.$$

That entropy condition introduces some irreversibility, and is expected to ensure the stability. Nevertheless, in dimension higher than 1, it is not sufficient to define a suitable mathematical framework, and to obtain a global existence and uniqueness theorem.

The only results about the existence of multidimensional discontinuous solutions known at the present time give the existence of shock profiles under some conditions (see [97], [98] or [47] for instance).

Regarding the uniqueness, as a byproduct of their analysis concerning weak solutions to the incompressible Euler equations, De Lellis and Székelyhidi [40] have obtained a negative result for some hyperbolic system of conservation laws, in particular for the p-system of isentropic gas dynamics

$$\begin{aligned} \partial_t R + \nabla_x \cdot (RU) &= 0, \\ \partial_t (RU) + \nabla_x \cdot (RU \otimes U + P(R)Id) &= 0. \end{aligned} \tag{6.4}$$

They have indeed proved that in dimension higher than 1, for any given function P, there exist bounded initial data (R_{in}, U_{in}) with $R_{in} \geq c > 0$ for which there are infinitely many bounded admissible solutions (R, U) of (6.4) with $R \geq c > 0$.

6.1.3 Global Entropic Solutions in One Spatial Dimension

In one spatial dimension, a good framework to study hyperbolic systems is the space of functions with bounded total variation, i.e. of L^∞ functions with derivatives in the space of bounded measures.

In that framework, if the hyperbolic system under consideration admits an entropy, then one has a theorem giving the global existence of entropic solutions for initial data with small total variation. In the particular case of the compressible Euler equations, the result can be stated as follows (see [78]).

Theorem 6.1.4 *Let R_0, T_0 be some nonnegative constants, and (R_{in}, U_{in}, T_{in}) be some perturbation with bounded variation of the constant state $(R_0, 0, T_0)$.*

Then if the total variation $TV(R_{in}, U_{in}, T_{in})$ is sufficiently small, the compressible Euler equations (6.1) admits a global entropic solution (R, U, T) on $\mathbf{R}^+ \times \mathbf{R}^3$.

Sketch of proof. That result is established by methods - based on numerical schemes - which describe approximatively the discontinuities and the way they propagate. The pioneering works in that direction are due to Glimm [50]. Alternative schemes have been given for instance by Bressan and Colombo [21].

• The first remark is that the Riemann problem, i.e. the Cauchy problem for initial data of the type

$$V_{in}(x) = V_- \text{ if } x < 0, \quad V_{in}(x) = V_+ \text{ if } x > 0$$

admits a solution (which can be computed almost explicitly) under the only condition that $\|V_+ - V_-\|_\infty$ is sufficiently small [67].

In order to obtain that solution, one has to study the *Rankine-Hugoniot curves*, i.e. the curves representing the states connected to V_- by a single wave (shock wave, or rarefaction wave, or contact discontinuity), then to prove that they provide a system of suitable coordinates, or in other words a good covering of the state space in the vicinity of V_-, using some local inversion theorem.

• The proof of the theorem relies then on four main arguments. One starts by building approximate solutions by gathering together elementary solutions of Riemann's problems. Then, for each one of these approximate solutions, one has to estimate the interactions between waves coming from two different Riemann's problems. These estimates on the *potential for interaction* allow to get compactness on that sequence of approximate solutions. The last step consists in proving that the limit points of this sequence are solutions to the original Cauchy problem.

• Actually the proof of Theorem 6.1.4 presents an additional difficulty related to the possible cavitation. Indeed, we recall that, when the density vanishes, the system (6.1) is no more hyperbolic so that the previous methods cannot be applied. One has therefore to check that the density remains bounded from below everywhere. That problem has been dealt with by Liu [78]. □

Note that the front tracking algorithm developed by Bressan and Colombo [21] (which is a variant of the previous approximation process) gives the uniqueness of solutions as well as their L^1 stability with respect to initial data (see [19]).

6.2 Some Perspectives

At the present time, the mathematical derivation of the compressible Euler equations from Boltzmann's kinetic theory remains an outstanding open problem.

The only rigorous results have been proved by Nishida [88] and Ukai and Asano [105], then improved by Liu and Yu [79], [80] and can be stated roughly as follows : as long as the compressible Euler equations have a smooth solution (R, U, T), one can construct a sequence of smooth solutions to the scaled Boltzmann equation the moments of which converge to (R, U, T). The method consists in deriving local strong a priori estimates (inherited from the propagation of regularity), and concluding by some fixed point argument, which is very similar to the proof of well-posedness for hyperbolic systems by Friedrichs [35]. Some works in progress by Métivier and Zumbrun [83] should further improve these results insofar as they also consider weak (viscous) shocks.

6.2.1 Convergence Towards Smooth Solutions

A first direction to extend that result would be to establish some strong-weak stability principle, i.e. to prove the convergence of any sequence of the appropriately scaled Boltzmann equation to the solution of the compressible Euler equations, as long as the latter does exist. In other words, this would be the counterpart of Theorem 5.1.10 that holds in incompressible regime.

A natural idea to do that is to derive some modulated entropy inequality, insofar as

- it involves only **physical quantities** (entropy, entropy dissipation, mass, momentum and energy);
- it looks very similar to the **strong-weak uniqueness** principle for symmetrizable hyperbolic systems (see [36] for instance).

As in the previous chapter, the difficulty will be to control the energy flux - which is defined by a moment of third order - in terms of the modulated entropy and entropy dissipation. This indeed requires to have a L^∞ bound with respect to the spatial variables, which is not known to hold for renormalized solutions to the Boltzmann equation.

Considering for instance the classical solutions built by Guo [62] allows to get rid of that problem, but it is then not clear that it improves the result by Liu, Yang and Yu [80].

6.2.2 Convergence Towards Weak Solutions in One Space Dimension

Another track would be to consider only the one dimensional case, for which the structure of hyperbolic systems is much better understood, exploiting in particular the similarities between the weak solutions to the Boltzmann equation built by Cercignani [29], and the entropic solutions to the compressible Euler system built by Glimm [50].

Indeed the key arguments leading to the global existence result in both cases are

- the **entropy inequality** choosing among elementary solutions the one which satisfies some causality principle;
- the a priori bound on some quantity, referred to as the **potential for interaction**, controlling the effect of the nonlinearity;
- the specificity of the **one dimensional** problem, due to the very particular properties of the scalar product.

Note that, in the absence of additional conditions, solutions of both equations are not known to be unique, so that we cannot expect to obtain a strong convergence statement.

The idea is therefore to obtain suitable a priori estimates on the moments of the scaled Boltzmann equation combining the bound on the potential for interaction together with some suitable micro-macro decomposition. The conclusion would then follow from some standard *moment method* (see the study of the incompressible Navier-Stokes asymptotics in Chapter 4).

Another approach would be to search for additional conditions which guarantee the uniqueness and some strong stability, such as the one given by Bressan and Colombo [21] for hyperbolic systems, which consists in restricting the attention to solutions obtained by front tracking methods.

The idea is therefore to define a similar algorithm of approximation for the solutions to the Boltzmann equation, and to prove that, for such solutions, there exists some *functional controlling the stability* and therefore the convergence to the compressible Euler equations.

Note that other attempts to obtain stability criteria for hyperbolic systems have been made for instance by Bardos and Pironneau [8] considering the linearized version of the equations in the vicinity of shocks. But for the moment we have no idea on possible use of these ideas in the context of hydrodynamic limits.

Appendix

A Some Consequences of Egorov's Theorem

Let us first recall the statement of Egorov's theorem :

Theorem A.1 *Let (X, μ) be some measure space with finite positive measure. Consider a sequence (g_n) such that $g_n \to g$ almost everywhere. Then, for each $\varepsilon > 0$, there exists a measurable $E \subset X$ such that*

$$\mu(X \setminus E) < \varepsilon \quad \text{and } g_n \to g \text{ uniformly on } E.$$

A.1 The Product Limit Theorem

We now give a corollary of Egorov's theorem, established by DiPerna and Lions [44], which is used repeatedly for the study of the Boltzmann equation and its hydrodynamic limits.

Proposition A.2 *Let (X, μ) be some measure space with finite positive measure. Consider two sequences of real-valued measurable functions defined on X, denoted (f_n) and (g_n). If (g_n) is bounded in $L^\infty(X)$ such that $g_n \to g$ almost everywhere, and $f_n \rightharpoonup f$ weakly in $L^1(X)$ then $f_n g_n \rightharpoonup fg$ weakly in $L^1(X)$.*

Proof. Without loss of generality, we can assume that $g = 0$.

Let δ be any fixed nonnegative constant. The sequence (f_n), being relatively weakly compact in $L^1(X)$, is equiintegrable. Thus, by picking $\alpha > 0$ sufficiently small, one has for every measurable set A such that $\mu(A) < \alpha$,

$$\int_A |f_n - f|(x) d\mu(x) < \delta \text{ uniformly in } n. \tag{A.1}$$

Fix such a constant α. By Egorov's theorem, as (g_n) is bounded in $L^\infty(X)$, and $g_n \to 0$ almost everywhere, there exists a measurable set A such that $\mu(A) < \alpha$ and

$$g_n \to 0 \text{ as } n \to \infty \text{ uniformly on } X \setminus A. \tag{A.2}$$

Fix such a set A. Then,

$$\int |(f_n - f)g_n|d\mu = \int_A |(f_n - f)g_n|d\mu + \int_{X\setminus A} |(f_n - f)g_n|d\mu$$

By (A.1) the first term in the right-hand side satisfies

$$\int_A |(f_n - f)g_n|d\mu \le \|g\|_{L^\infty(X)} \int_A |f_n - f|d\mu \le \|g\|_{L^\infty(X)}\delta,$$

whereas by (A.2) the second term converges to 0 as $n \to \infty$:

$$\int_{X\setminus A} |(f_n - f)g_n|d\mu \le \left(\sup_n \|f_n - f\|_{L^1(X)} \right) \|g_n\|_{L^\infty(X\setminus A)} \to 0.$$

Finally

$$\lim_{n\to\infty} \int |(f_n - f)g_n|d\mu \le \|g\|_{L^\infty(X)}\delta,$$

but δ was arbitrary, whence Proposition A.2 holds. $\qquad\square$

A.2 An Asymptotic Result of Variables Separating

For the study of boundary conditions, we also need the following variant of the Product Limit theorem, which has been proved in [82] :

Proposition A.3 *Let (X, μ_X) and (Y, μ_Y) be two measure spaces of finite measures. Consider a family of nonnegative functions (χ_ε) uniformly bounded in $L^\infty(X \times Y)$ converging almost everywhere to 1 on $X \times Y$, and a family (ρ_ε) of nonnegative functions of $L^1(X)$ such that*

$$(\chi_\varepsilon\rho_\varepsilon) \text{ is relatively weakly compact in } L^1(X \times Y)$$

Then any limit point ρ of $(\chi_\varepsilon\rho_\varepsilon)$ belongs to $L^1(X)$, namely does not depend on $y \in Y$.

Before giving the proof, let us notice that if (ρ_ε) were supposed to be relatively weakly compact in $L^1(X \times Y)$ then the conclusion of the lemma would be straightforward.

Proof. Consider a subsequence of $(\chi_\varepsilon\rho_\varepsilon)$ (still denoted $(\chi_\varepsilon\rho_\varepsilon)$) converging to ρ as $\varepsilon \to 0$. Let δ be any fixed nonnegative constant. As χ_ε converges almost everywhere to 1, there exist $A \subset X \times Y$ with $|X \times Y \setminus A| \le \delta$, and $\varepsilon_0 > 0$ such that

$$\forall \varepsilon \le \varepsilon_0, \quad \chi_{\varepsilon|A} \ge \frac{1}{2}$$

By the Product Limit theorem, as $\left(\dfrac{1_A}{\chi_\varepsilon}\right)_{\varepsilon \le \varepsilon_0}$ is bounded in $L^\infty(X \times Y)$ and converges a.e. to 1_A,

$$\rho_\varepsilon 1_A = \rho_\varepsilon \chi_\varepsilon \frac{1_A}{\chi_\varepsilon} \rightharpoonup \rho 1_A \text{ weakly in } L^1(X \times Y).$$

Define

$$A_X = \left\{ x \in X \; / \; \int 1_A(x,y) d\mu_Y(y) \ge \frac{\mu_Y(Y)}{2} \right\}.$$

From Bienaymé Tchebichev's inequality

$$1_{A_X}(x)\rho_\varepsilon(x) \le 1_{A_X}(x)\rho_\varepsilon(x) \frac{2 \int 1_A(x,y) d\mu_Y(y)}{\mu_Y(Y)}$$
$$= \frac{2 \, 1_{A_X}}{\mu_Y(Y)} \int \rho_\varepsilon(x) 1_A(x,y) d\mu_Y(y) \tag{A.3}$$

we deduce that $(1_{A_X}\rho_\varepsilon)$ is weakly compact in $L^1(X)$. Then, up to extraction of a subsequence,

$$1_{A_X}\rho_\varepsilon \rightharpoonup \tilde\rho \text{ weakly in } L^1(X)$$

By the Product Limit theorem, as (χ_ε) is bounded in $L^\infty(X \times Y)$ and converges a.e. to 1,

$$1_{A_X}\rho_\varepsilon \chi_\varepsilon \rightharpoonup \tilde\rho \text{ weakly in } L^1(X \times Y)$$

from which we deduce that

$$\rho 1_{A_X} = \tilde\rho \in L^1(X) \tag{A.4}$$

On the other hand, if $x \notin A_X$,

$$\int 1_{X \times Y \setminus A}(x,y) d\mu_Y(y) = \int (1 - 1_A)(x,y) d\mu_Y(y) \ge \frac{\mu_Y(Y)}{2}$$

Then,

$$\int 1_{X \setminus A_X}(x) d\mu_X(x) \le \int 1_{X \setminus A_X}(x) \frac{2 \int 1_{X \times Y \setminus A}(x,y) d\mu_Y(y)}{\mu_Y(Y)} d\mu_X(x)$$
$$\le \frac{2}{\mu_Y(Y)} \delta \tag{A.5}$$

As there exists A_X satisfying (A.4) and (A.5) for all $\delta > 0$, ρ depends only on the variable x, and thus $\rho \in L^1(X)$.   □

B Classical Trace Results on the Solutions of Transport Equations

In order to deal with kinetic equations involving reflection conditions at the boundary, we need the following fundamental result due to Cessenat [32] following Bardos [3] and Ukai [104], which allows to define the trace of any weak solution to the free-transport equation in a very general setting.

B.1 Definition of the Trace

Proposition B.1 *For any smooth subset Ω of $\mathbf{R}^3$, denote by $W^p(\Omega)$ the functional space*

$$\{f \in L^p(\mathbf{R} \times \Omega \times \mathbf{R}^3) \,/\, (\mathrm{St}\partial_t + v \cdot \nabla_x)f \in L^p(\mathbf{R} \times \Omega \times \mathbf{R}^3)\}.$$

Then the trace operator

$$\gamma : f \in W^p(\Omega) \mapsto f_{|\partial\Omega} \in L^p(\mathbf{R} \times \partial\Omega \times \mathbf{R}^3, dt d\sigma_x |v \cdot n(x)|^2 (1 + |v|)^{-1} dv)$$

is continuous.

Proof. Without loss of generality, we can restrict our attention to nonnegative functions.

By Green's formula, we have for any bounded function $\varphi \in C^1(\bar{\Omega} \times \mathbf{R}^3)$,

$$
\begin{aligned}
p &\iiint \varphi(x,v) f^{p-1} (\mathrm{St}\partial_t + v \cdot \nabla_x) f(t,x,v) dt dx dv \\
&+ \iiint (v \cdot \nabla_x) \varphi(x,v) f^p(t,x,v) dt dx dv \\
&= \iiint \varphi(x,v) f^p(t,x,v)(v \cdot n(x)) dt d\sigma_x dv
\end{aligned}
\tag{B.6}
$$

As Ω is assumed to be smooth, there exists some vector field $n \in W^{1,\infty}(\bar{\Omega})$ which coincides with the outward unit normal at the boundary. Thus, choosing

$$\varphi(x,v) = \frac{(v \cdot n(x))}{(1 + |v|^2)^{1/2}}$$

we get

$$
\begin{aligned}
\|f_{|\partial\Omega}\|_{L^p(dt d\sigma_x |v \cdot n(x)|^2 (1+|v|)^{-1} dv)} & \\
\leq C \left(\|f\|_{L^p(dt dx dv)} + \|(\mathrm{St} + v \cdot \nabla_x)f\|_{L^p(dt dx dv)} \right),
\end{aligned}
\tag{B.7}
$$

which concludes the proof.   □

B.2 Free Transport with Reflection at the Boundary

In order to extend regularity and dispersion results to free transport with reflection at the boundary, we have then to establish a priori estimates on the incoming flux (which is defined in terms of the outgoing flux).

Proposition B.2 *Let Ω be any smooth subset of $\mathbf{R}^3$, and $f \in L^1(\mathbf{R} \times \Omega \times \mathbf{R}^3)$ be a solution to the free-transport equation*

$$\mathrm{St}\partial_t f + v \cdot \nabla_x f = S$$

supplemented with Maxwell's boundary condition

$$f_{|\Sigma_-} = (1 - \alpha)Lf_{|\Sigma_+} + \alpha Kf_{|\Sigma_+} \quad on \ \Sigma_-$$

where the outgoing/incoming sets Σ_+ and Σ_- at the boundary $\partial\Omega$ are defined by

$$\Sigma_{\pm} = \{(x, v) \in \partial\Omega \times \mathbf{R}^3, \quad \pm n(x) \cdot v > 0\},$$

the local reflection operator L is given by

$$Lf(x, v) = f(x, v - 2(v \cdot n(x))n(x)),$$

and the diffuse reflection operator K is given by

$$Kf(x, v) = M_w(v) \int_{v' . n(x) > 0} f(x, v') (v' \cdot n(x)) dv'$$

for some normalized Maxwellian distribution M_w characterizing the state of the wall.

Then, there exists some nonnegative constant C (depending on the Lipschitz norm of n) such that

$$\iiint f_{|\Sigma} |v \cdot n(x)| dv dx dt \leq \frac{C}{\alpha} (\|f\|_{L^1(\mathbf{R} \times \Omega \times \mathbf{R}^3)} + \|S\|_{L^1(\mathbf{R} \times \Omega \times \mathbf{R}^3)}).$$

Proof. The smoothness assumption made on the boundary implies the existence of a vector field n which belongs to $W^{1,\infty}(\bar{\Omega})$ and coincides with the outward unit normal vector at the boundary. Therefore, multiplying the transport equation by $(n(x) \cdot v)/(1+|v|)$ and integrating with respect to all variables, we get

$$\iiint \frac{v \cdot n(x)}{1 + |v|} (v \cdot \nabla_x) f \, dv dx dt = \iiint \frac{v \cdot n(x)}{1 + |v|} S \, dv dx dt$$

Then, using Green's formula, we get

$$\iiint f_{|\Sigma} \frac{(v \cdot n(x))^2}{1 + |v|} dv d\sigma_x dt \leq C(\|f\|_{L^1(\mathbf{R} \times \Omega \times \mathbf{R}^3)} + \|S\|_{L^1(\mathbf{R} \times \Omega \times \mathbf{R}^3)}).$$

In particular

$$\alpha \iiint_{v \cdot n < 0} Kf_{|\Sigma_+} \frac{(v \cdot n(x))^2}{1 + |v|} dv d\sigma_x dt \leq C(\|f\|_{L^1(\mathbf{R} \times \Omega \times \mathbf{R}^3)} + \|S\|_{L^1(\mathbf{R} \times \Omega \times \mathbf{R}^3)}).$$

By definition of K, we have the spreading condition

$$\int_{v \cdot n(x) > 0} f_{|\Sigma_+} (v \cdot n(x)) dv \leq \kappa_0 \int_{v \cdot n(x) < 0} Kf_{|\Sigma_+} \frac{(v \cdot n(x))^2}{1 + |v|} dv.$$

We then deduce that

$$\iiint_{v \cdot n(x) > 0} f_{|\Sigma_+} (v \cdot n(x)) dv d\sigma_x dt \leq \frac{C\kappa_0}{\alpha} (\|f\|_{L^1(\mathbf{R} \times \Omega \times \mathbf{R}^3)} + \|S\|_{L^1(\mathbf{R} \times \Omega \times \mathbf{R}^3)}).$$

On the other hand, the normalization condition on M_w implies

$$-\int_{v\cdot n(x)<0} f_{|\Sigma}(v\cdot n(x))dv = \int_{v\cdot n(x)>0} f_{|\Sigma}(v\cdot n(x))dv.$$

We therefore have

$$\iiint f_{|\Sigma}|v\cdot n(x)|dvd\sigma_x dt \leq \frac{2C\kappa_0}{\alpha}(\|f\|_{L^1(\mathbf{R}\times\Omega\times\mathbf{R}^3)} + \|S\|_{L^1(\mathbf{R}\times\Omega\times\mathbf{R}^3)}).$$

which is the expected inequality.   □

Remark B.3 *Note that, in the case of a purely specular reflection, we do not obtain such a bound on the trace. Nevertheless, because the specular reflection is completely transparent in the weak formulation of the transport equation (by obvious symmetry properties), the study is actually much easier (very similar to the case when there is no boundary). We refer to the work [63] by Hamdache for a careful treatment of that case.*

C Some Consequences of Chacon's Biting Lemma

Let us first recall the statement of Chacon's Biting Lemma [21]:

Theorem C.1 *Let (X,μ) be some measure space with finite positive measure. Consider a sequence (g_n) bounded in $L^1(X)$. Then, there exists a subsequence (g'_n) of (g_n) and some function $g \in L^1(X)$ such that $g'_n \to g$ in the sense of Chacon, meaning that for each $\varepsilon > 0$, there exists a measurable $E \subset X$ such that $\mu(X \setminus E) \leq \varepsilon$ and*

$$g'_n \to g \text{ in } L^1(E).$$

C.1 From Renormalized Convergence to Chacon's Convergence

We now give an extension of Chacon's Biting Lemma to the space of measurable and almost everywhere finite functions, established by Mischler [85] to study the traces of kinetic equations in the framework of renormalized solutions.

This requires the following definition of *renormalized convergence* (see [85] and the references therein for basic results concerning renormalized convergence) :

Definition C.2 *A sequence (g_n) of measurable and almost everywhere finite functions is said to converge in renormalized sense to some measurable and almost everywhere finite function g, if for any increasing sequence $\Gamma_M \in C \cap L^\infty(\mathbf{R}^+)$ converging simply to $\mathrm{Id}_{|\mathbf{R}^+}$ as $M \to \infty$, and any subsequence (g'_n) of (g_n), there exists a sequence γ_M and a subsequence (g''_n) of (g'_n) such that*

$$\Gamma_M(g''_n) \rightharpoonup \gamma_M \text{ weakly-}* \text{ in } L^\infty(X) \text{ and } \gamma_M \to g \text{ a.e. in } X.$$

We have then the following relation between renormalized convergence and convergence in the sense of Chacon (see [85]) :

Proposition C.3 *Let* (X, μ) *be some measure space with finite positive measure. Consider a sequence* (g_n) *of measurable and almost everywhere finite functions, such that*

$$g_n \to g \text{ in renormalized sense,}$$

where g *is some measurable and almost everywhere finite function. Then,*

$$\lim_{M \to +\infty} \sup_n \mu(\{g_n \geq M\}) = 0,$$

and there exists a subsequence (g'_n) *of* (g_n) *such that* $g'_n \to g$ *in the sense of Chacon.*

Proof. Without loss of generality, we can restrict our attention to nonnegative functions.

- We first prove the L^0 bound, arguing by contradiction. Since g is a measurable and almost everywhere finite function, for any arbitrary $\varepsilon > 0$, there exists $E \subset X$ such that $\mu(X \setminus E) < \varepsilon$ and $g \in L^1(E)$.

If there is no m such that

$$\sup_n \mu(\{g_n \geq m\}) < \varepsilon$$

there exists an increasing sequence (n_m) such that

$$\forall m \in \mathbf{N}, \quad \mu(\{g_{n_m} \geq m\}) \geq \varepsilon.$$

Therefore, for any $l \in \mathbf{N}$ and any $m \geq l$,

$$\int_E \Gamma_l(g_{n_m}) d\mu \geq l\varepsilon,$$

where Γ_l is some smooth version of the truncation $x \mapsto \min(x, l)$. Passing to the limit $m \to \infty$ leads to

$$\int_E g d\mu \geq \int_E \gamma_l d\mu \geq \varepsilon l,$$

by definition of the renormalized convergence. Thus

$$\int_E g d\mu \geq \lim_{l \to \infty} \varepsilon l,$$

which gives the expected contradiction.

- We have then to prove the convergence in the sense of Chacon. Given $\varepsilon > 0$, one can choose $E \subset X$ such that $\mu(X \setminus E) < \varepsilon$ and $g \in L^1(E)$.

We construct a first subsequence (n_l) such that

$$\int_E \Gamma_l(g_{n_l})d\mu \leq \int_E g d\mu + \frac{1}{l}.$$

Then, for any $m \leq l$, $\Gamma_m(g_{n_l}) \leq \Gamma_l(g_{n_l})$ so that

$$g = \limsup_{m\to\infty} \gamma_m \leq \limsup_{m\to\infty} \liminf_{l\to\infty} \Gamma_l(g_{n_l}),$$

where the lim inf is taken in the sense of Chacon. By Fatou's lemma and the definition of the renormalized convergence, we also have

$$\forall A \subset X, \quad \int_A \limsup_{l\to\infty} \Gamma_l(g_{n_l})d\mu \leq \limsup_{l\to\infty} \int_A \Gamma_l(g_{n_l})d\mu \leq \int_A g d\mu.$$

Combining all these results show that

$$\Gamma_l(g_{n_l}) \to g \text{ in the sense of Chacon on } E.$$

In other words, there exists E' such that $\mu(E \setminus E') < \varepsilon$ and

$$\Gamma_l(g_{n_l}) \rightharpoonup g \text{ weakly in } L^1(E').$$

Furthermore, since (g_n) is bounded in L^0, one can choose a second subsequence (still denoted (g_{n_l})) such that the sets Z_L defined by

$$Z_L = \{\exists l \geq L \, / \, g_{n_l} \neq \Gamma_l(g_{n_l})\}$$

satisfy

$$\mu(Z_L) \leq \sum_{l \geq L} \mu(\{g_{n_l} > l\}) \to 0 \text{ as } L \to \infty.$$

Finally, choosing L large enough such that $\mu(Z_L) < \varepsilon$, and setting $E'' = E' \setminus Z_L$, we obtain

$$\mu(X \setminus E'') < 3\varepsilon \text{ and } g_{n_l} \rightharpoonup g \text{ weakly in } L^1(E'').$$

We conclude thanks to a diagonal process.   □

C.2 A Result of Partial Equiintegrability

In order to characterize the limiting incoming flux in Maxwell's boundary condition, we also need the following variant of the previous result, also established in [85].

Proposition C.4 *Let (X, μ_X) be some measure space with finite positive measure, and $(\mu_Y(x))_{x \in X}$ a family of probability measures on the space Y, and denote $\langle g \rangle_Y = \int g d\mu_Y$.*

Let $h \in C(\mathbf{R}^+, \mathbf{R}^+)$ be some convex function of class $C^2(\mathbf{R}_*^+)$ with superlinear growth at infinity, and such that $H(s,t) \equiv (h(t) - h(s))(t - s)$ is convex.

Consider a sequence (g_n) of measurable nonnegative and almost everywhere finite functions on $X \times Y$, such that

$$\int \left(\langle h(g_n) \rangle_Y - h(\langle g_n \rangle_Y) \right) d\mu_X \leq C, \tag{C.8}$$

$$\langle g_n \rangle_Y \to \bar{g} \text{ in renormalized sense on } X$$

Then there exists $g \in L^1(X \times Y, d\mu_X(x)d\mu_Y(x,y))$ and a subsequence (g_n') of (g_n) such that, for every $\varepsilon > 0$, one can find some $E \subset X$ with

$$\mu_X(X \setminus E) < \varepsilon \text{ and } g_n' \rightharpoonup g \text{ weakly in } L^1(E \times Y).$$

In particular, $\bar{g}(x) = \langle g \rangle_Y$ almost everywhere.

Furthermore

$$\int \left(\langle h(g) \rangle_Y - h(\langle g \rangle_Y) \right) d\mu_X \leq C.$$

Proof. The point to be understood here is how the (convex) functional which generalizes the Darrozès-Guiraud information allows to gain some equiintegrability with respect to y, and thus to establish the convergence of some integral quantities.

• By Proposition C.3, we deduce from the renormalized convergence in (C.8) that, up to extraction of a subsequence,

$$\langle g_n \rangle_Y \to \bar{g} \text{ in the sense of Chacon.}$$

In particular, for any $\varepsilon > 0$, there exists $A \subset X$ such that

$$\mu_X(X \setminus A) < \varepsilon \quad \text{and} \quad \langle g_n \rangle_Y \to \bar{g} \text{ weakly in } L^1(A).$$

Thanks to Dunford-Pettis' lemma, there is therefore a (nonnegative increasing) convex function Φ with superlinear growth at infinity such that $\Phi(0) = 0$, $\Phi'(0) > 0$ and

$$\int_A \Phi(\langle g_n \rangle_Y) d\mu_X \leq C_1.$$

We are then able to build some (nonnegative increasing) convex function Ψ with superlinear growth at infinity such that $\Psi(0) = 0$, $\Psi'(0) > 0$ and

$$\Psi \leq \Phi,$$
$$h - \Psi \text{ is convex.}$$

Jensen's inequality, written for the function $h - \Psi$, combined with the uniform bound in (C.8), gives therefore

$$\int \Big(\langle \Psi(g_n) \rangle_Y - \Psi \big(\langle g_n \rangle_Y \big) \Big) d\mu_X \leq C,$$

and thus

$$\iint_{A \times Y} \Psi(g_n) d\mu_Y d\mu_X \leq C + C_1.$$

By Dunford-Pettis' lemma, we obtain that (g_n) belongs to a weak compact subset of $L^1(A \times Y)$.

We conclude, by a diagonal process, that there is a function g in $L^1(X \times Y)$ and a subsequence of (g_n) which converges to g in the sense stated in Proposition C.4. In particular, for any $\varepsilon > 0$, there exists $A \subset X$ such that

$$\mu_X(X \setminus A) < \varepsilon \quad \text{and} \quad \langle g_n \rangle_Y \to \langle g \rangle_Y \text{ in } L^1(A).$$

Identifying the limit leads to $\bar{g} = \langle g \rangle_Y$ for almost every $x \in X$.

- It remains then to take limits in the uniform bound in (C.8).

We start by proving that, for all $x \in X$,

$$E : g \in L^1(Y) \mapsto E(g) = \Big\langle h(g) \Big\rangle_Y - h \Big(\langle g(y) \rangle_Y \Big)$$

is a convex functional. We proceed indeed by approximation replacing h by $h_\varepsilon : z \mapsto h(z + \varepsilon) - h(\varepsilon)$. As $h_\varepsilon \in C^2(\mathbf{R}^+)$,

$$DE_\varepsilon(g_1) \cdot g_2 = \Big\langle h'_\varepsilon(g_1).g_2 \Big\rangle_Y - h'_\varepsilon \Big(\langle g_1 \rangle_Y \Big) \langle g_2 \rangle_Y.$$

Therefore, by Jensen's inequality, we have for $g_1, g_2 \in L^\infty(Y)$

$$\Big(DE_\varepsilon(g_1) - DE_\varepsilon(g_2) \Big) \cdot (g_1 - g_2)$$
$$= \Big\langle H(g_1, g_2) \Big\rangle_Y - H \Big(\langle g_1 \rangle_Y, \langle g_2 \rangle_Y \Big) \geq 0,$$

so that DE_ε is monotone and E_ε is convex on $L^\infty(Y)$. Passing to the limit $\varepsilon \to 0$, we then obtain that E is convex on $L^\infty(Y)$:

$$\forall t \in [0, 1], \quad E(tg_1 + (1 - t)g_2)) \leq tE(g_1) + (1 - t)E(g_2).$$

Now let $g_1, g_2 \in L^1(Y)$. If $h(g_1)$ or $h(g_2)$ does not belong to $L^1(Y)$ the convex inequality obviously holds. In the other case, we choose two sequences $(g_{1\varepsilon})$ and $(g_{2\varepsilon})$ of $L^\infty(Y)$ such that $g_{1\varepsilon} \nearrow g_1$ and $g_{2\varepsilon} \nearrow g_2$ almost everywhere, and passing to the limit $\varepsilon \to 0$ in the convex inequality written for $g_{1\varepsilon}$ and $g_{2\varepsilon}$, we get by Lebesgue's theorem and Fatou's lemma that

$$\forall t \in [0, 1], \quad E(tg_1 + (1 - t)g_2)) \leq tE(g_1) + (1 - t)E(g_2).$$

Now, if $g_1, g_2 \in L^1(X \times Y)$, then $g_1(x, .), g_2(x, .) \in L^1(Y)$ for almost all $x \in X$ and, integrating the previous convex inequality, we obtain that the functional

$$g \in L^1(X \times Y) \mapsto \int E(g)d\mu_X$$

is convex. Furthermore, by Fatou's Lemma, this functional is lower semi-continuous.

From the convergence stated in Proposition C.4 and established previously, we then deduce that

$$\int \left(\langle h(g) \rangle_Y - h\left(\langle g \rangle_Y\right) \right) d\mu_X \leq C,$$

which concludes the proof.   □

References

1. V.I. Agoshkov: *Spaces of functions with differential difference characteristics and smoothness of the solutions of the transport equations*, Dokl. Akad. Nauk SSSR **276** (1984), 1289–1293.
2. J.-P. Aubin: *Un théorème de compacité.*, C. R. Acad. Sci. Paris **256** (1963), 5042–5044.
3. C. Bardos : *Problèmes aux limites pour les équations aux dérivées partielles du premier ordre à coefficients réels; théorèmes d'approximation; application à l'équation de transport*, Ann. Scient. Ec. Norm. Sup. **3** (1970), 185–233.
4. C. Bardos, F. Golse, C.D. Levermore: *Fluid Dynamic Limits of the Boltzmann Equation I*, J. Stat, Phys. **63** (1991), 323–344.
5. C. Bardos, F. Golse, C.D. Levermore: *Fluid Dynamic Limits of Kinetic Equations II: Convergence Proofs for the Boltzmann Equation*, Comm. Pure & Appl. Math **46** (1993), 667–753.
6. C. Bardos, F. Golse, C.D. Levermore. *The acoustic limit for the Boltzmann equation*. Arch. Ration. Mech. Anal. **153** (2000), 177–204.
7. C. Bardos, S. Ukai: *The classical incompressible Navier-Stokes limit of the Boltzmann equation*, Math. Models and Methods in the Appl. Sci. **1** (1991), 235–257.
8. C. Bardos, O. Pironneau. *A formalism for the differentiation of conservation laws*, C. R. Math. Acad. Sci. Paris **335** (2002), 839–845.
9. C. Bardos, E. Titi: *Euler Equations of Incompressible Ideal Fluids*, Preprint (2008).
10. J.T. Beale, T. Kato, A.J. Majda: *Remarks on the breakdown of smooth solutions for the 3D Euler equations*, Commun. Math. Phys. **1994** (1984), 61–66.
11. A. Bobylev. *The theory of the nonlinear, spatially uniform Boltzmann equation for Maxwellian molecules*, Sov. Sci. Rev. C. Math. Phys. **7** (1998), 111–233.
12. A. Bobylev, C. Cercignani. *On the rate of entropy production for the Boltzmann equation*, J. Statis. Phys **94** (1999), 603–618.
13. A. Bobylev, D. Levermore., in preparation.
14. L. Boltzmann. *Über die Prinzipien der Mechanik: Zwei Akademische Antrittsreden*. Leipzig : S. Hirzel, 1903.
15. F. Bouchut, L. Desvillettes. *A proof of the smoothing properties of the positive part of Boltzmann's kernel*, Rev. Mat. Iberoamericana **14** (1998), 47–61.

182   References

16. F. Bouchut, F. Golse, M. Pulvirenti: "Kinetic Equations and Asymptotic Theory",
    L. Desvillettes & B. Perthame ed., Editions scientifiques et médicales Elsevier,
    Paris, 2000.
17. Y. Brenier. *Convergence of the Vlasov-Poisson system to the incompressible Euler
    equations,* Comm. Partial Differential Equations **25** (2000), 737–754.
18. A. Bressan, R. Colombo. *The semigroup generated by* 2×2 *conservation laws,*
    Arch. Rational Mech. Anal. **133** (1995), 1–75.
19. A. Bressan, S. Bianchini. *Vanishing viscosity solutions of nonlinear hyperbolic
    systems,* Ann. of Math. **161** (2005), 223–342.
20. H. Brézis, J.P. Bourguignon. *Remarks on the Euler equation.* J. Functional Anal-
    ysis **15** (1974), 341–363.
21. J. Brooks, R. Chacon. *Continuity and compactness of measures.* Adv. in Math.
    37 (1980), 16–26.
22. L. Caffarelli, R. Kohn, L. Nirenberg. *Partial regularity of suitable weak solutions
    of the Navier-Stokes equations,* Comm. Pure Appl. Math. **35** (1982), 771–831.
23. R.E. Caflisch: *The Boltzmann equation with a soft potential. I. Linear, spatially-
    homogeneous,* Commun. Math. Phys. **74** (1980), 71–95.
24. R.E. Caflisch: *The fluid dynamic limit of the nonlinear Boltzmann equation,*
    Comm. on Pure and Appl. Math. **33** (1980), 651–666.
25. M. Cannone, Y. Meyer, F. Planchon. *Solutions auto-similaires des équations de
    Navier-Stokes,* Séminaire sur les Equations aux Dérivées Partielles **8**, Ecole Poly-
    tech., Palaiseau, 1994.
26. F. Castella, B. Perthame. *Estimations de Strichartz pour les équations de trans-
    port cinétique,* C. R. Acad. Sci. Paris Sr. I Math. **322** (1996), 535–540.
27. J.-Y. Chemin. *Fluides parfaits incompressibles [Incompressible perfect fluids],*
    Astérisque **230**, 1995.
28. C. Cercignani: "The Boltzmann Equation and Its Applications" Springer-Verlag,
    New-York NY, 1988.
29. C. Cercignani. *Global weak solutions of the Boltzmann equation,* J. Stat. Phys.
    118 (2005), 333–342.
30. C. Cercignani, R. Illner. *Global weak solutions to the Boltzmann equation in a
    slab with diffusive boundary conditions,* Arch. Rational Mech. Anal. **134** (1996),
    1–16.
31. C. Cercignani, R. Illner, M. Pulvirenti, *The Mathematical Theory of Dilute Gases,*
    Springer Verlag, New York NY, 1994.
32. M. Cessenat. *Théorèmes de traces pour des espaces de fonctions de la neutronique,*
    C. R. Acad. Sci. Paris **300** (1985), 89–92.
33. S. Chapman, T.G. Cowling: "The mathematical theory of non-uniform gases: An
    account of the kinetic theory of viscosity, thermal conduction, and diffusion in
    gases". Cambridge University Press, New York, 1960.
34. P. Constantin, C. Foias: "Navier-Stokes equations." Chicago Lectures in Mathe-
    matics. University of Chicago Press, Chicago, 1988.
35. R. Courant, K.O. Friedrichs: "Supersonic flow and shock waves." Springer-Verlag,
    New York-Heidelberg, 1976.
36. C.M. Dafermos: "Hyperbolic conservation laws in continuum physics."
    Grundlehren der Mathematischen Wissenschaften **325**, Springer-Verlag, Berlin,
    2000.
37. Darrozès, J.S.; Guiraud, J.P.. *Généralisation formelle du théorème H en présence
    de parois,* C.R. Acad. Sci. Paris **262** (1966), 368–371.

38. Dautray, R.; Lions, J.L.. "Analyse mathématique et calcul numérique pour les sciences et les techniques" **9**, Masson, Paris, 1988.

39. C. De Lellis: *Notes on hyperbolic systems of conservation laws and transport equations*, Preprint (2006).

40. C. De Lellis, L. Székelyhidi: *On admissibility criteria for weak solutions of the Euler equations*, Preprint (2008).

41. J.-M. Delort. *Existence de nappes de tourbillon en dimension deux [Existence of vortex sheets in dimension two]*, J. Amer. Math. Soc. **4** (1991), 553–586.

42. A. DeMasi, R. Esposito, J. Lebowitz, *Incompressible Navier-Stokes and Euler Limits of the Boltzmann Equation,* Comm Pure Appl. Math. **42** (1990), 1189–1214.

43. L. Desvillettes, F. Golse, *A remark concerning the Chapman-Enskog asymptotics*, in "Advances in kinetic theory and computing", B. Perthame ed., 191–203, Ser. Adv. Math. Appl. Sci., 22, World Sci. Publishing, River Edge, NJ, 1994.

44. R.J. DiPerna, P.-L. Lions: *On the Cauchy problem for the Boltzmann equation: global existence and weak stability results*, Ann. of Math. **130** (1990), 321–366.

45. R. Ellis & M. Pinsky. *The first and second fluid approximations to the linearized Boltzmann equation,*. J. Math. Pures Appl. **54** (1975), 125–156.

46. R. Esposito, R. Marra, H.T. Yau. *Navier-Stokes equations for stochastic particle systems on the lattice.* Comm. Math. Phys. **182** (1996), 395–456.

47. J. Francheteau & G. Métivier: *Existence de chocs faibles pour des systmes quasilinaires hyperboliques multidimensionnels [Existence of weak shocks for multidimensional hyperbolic quasilinear systems]*, Astérisque **268** (2000).

48. H. Fujita, T. Kato. *On the Navier-Stokes initial value problem. I.* Arch. Rational Mech. Anal. **16** (1964), 269–315.

49. T. Gallay, C.E. Wayne, *Invariant manifolds and the long-time asymptotics of the Navier-Stokes and vorticity equations on* $\mathbf{R}^2$, Arch. Rat. Mech. Anal. **163** (2002), 209–258.

50. J. Glimm: *Solutions in the large for nonlinear hyperbolic systems of equations*, Comm. on Pure and Appl. Math. **18** (1965), 697–715.

51. F. Golse, D. Levermore. *The Stokes-Fourier and Acoustic Limits for the Boltzmann Equation*, Comm. on Pure and Appl. Math. **55** (2002), 336–393.

52. F. Golse, D. Levermore, L. Saint-Raymond. *La méthode de l'entropie relative pour les limites hydrodynamiques de modèles cinétiques* Séminaire Equations aux dérivées partielles (Polytechnique) (1999-2000).

53. F. Golse, P.-L. Lions, B. Perthame, R. Sentis: *Regularity of the moments of the solution of a transport equation*, J. Funct. Anal. **76** (1988), 110–125.

54. F. Golse, L. Saint-Raymond. *The Navier-Stokes limit of the Boltzmann equation for bounded collision kernels*, Invent. Math. **155** (2004), no. 1, 81–161.

55. F. Golse, L. Saint-Raymond. *The Navier-Stokes limit of the Boltzmann equation for hard potentials*, to appear in J. Math. Pures Appl. (2008).

56. F. Golse, L. Saint-Raymond: *Velocity averaging in* L^1 *for the transport equation*, C. R. Acad. Sci. **334** (2002), 557–562.

57. F. Golse, L. Saint-Raymond. *A remark about the asymptotic theory of the Boltzmann equation*, preprint.

58. F. Golse, L. Saint-Raymond. *Hydrodynamic limits for the Boltzmann equation*, Riv. Mat. Univ. Parma **4** (2005), 1–144.

59. H. Grad. *Asymptotic theory of the Boltzmann equation II* Rarefied Gas Dynamics (Proc. of the 3rd Intern. Sympos. Palais de l'UNESCO, Paris, 1962) Vol. I, 26–59.

60. E. Grenier. *Quelques limites singulières oscillantes*, Séminaire sur les Equations aux Dérivées Partielles **21**, Ecole Polytech., Palaiseau, 1995.
61. E. Grenier. *On the nonlinear instability of Euler and Prandtl equations*, Comm. Pure Appl. Math. **53** (2000), 1067–1091.
62. Y. Guo. *The Boltzmann equation in the whole space*, Indiana Univ. Math. J. **53** (2004), 1081–1094.
63. K. Hamdache: *Problèmes aux limites pour l'équation de Boltzmann: existence globale de solutions [Boundary value problems for the Boltzmann equation: global existence of solutions]*, Comm. Partial Differential Equations **13** (1988), 813–845.
64. H. Hertz. *Die Principien der Mechanik.* Leipzig, 1894.
65. D. Hilbert. *Begründung der kinetischen Gastheorie* Math. Ann. **72** (1912), 562–577.
66. M. Lachowicz, *On the initial layer and the existence theorem for the nonlinear Boltzmann equation.* Math. Methods Appl. Sci. **9** (1987), 342–366.
67. P.D. Lax: *Hyperbolic systems of conservation laws. II*, Comm. Pure Appl. Math. **10** (1957), 537–566.
68. C. Landim, H.T. Yau. *Fluctuation-dissipation equation of asymmetric simple exclusion processes.* Probab. Theory Related Fields **108** (1997), 321–356.
69. O.E. Lanford, *Time evolution of large classical systems* Lect. Notes in Physics **38**, J. Moser ed., 1–111, Springer Verlag (1975).
70. J. Leray. *Etude de diverses équations intégrales non linéaires et quelques problèmes que pose l'hydrodynamique*, J. Math. Pures Appl. **9** (1933), 1–82.
71. L. Lichtenstein: *Über einige Existenz Problem der hydrodynamik homogener unzusammendrückbarer, reibunglosser Flüssikeiten und die Helmholtzschen Wirbelsalitze*, Mat. Zeit. Phys. **23** (1925), 89154; **26** (1927), 193323; **32** (1930), 608.
72. P.-L. Lions, *Compactness in Boltzmann's equation via Fourier integral operators and applications* J. Math. Kyoto Univ. **34** (1994), 391–427, 429–461, 539–584.
73. P.-L. Lions, *Conditions at infinity for Boltzmann's equation*, Comm. Partial Differential Equations **19** (1994), 335–367.
74. P.-L. Lions: "Mathematical Topics in Fluid Mechanics, Vol. 1: Incompressible Models", The Clarendon Press, Oxford University Press, New York, 1996.
75. P.-L. Lions, N. Masmoudi: *From Boltzmann Equation to the Navier-Stokes and Euler Equations I*, Archive Rat. Mech. & Anal. **158** (2001), 173–193.
76. P.-L. Lions, N. Masmoudi: *From Boltzmann Equation to the Navier-Stokes and Euler Equations II*, Archive Rat. Mech. & Anal. **158** (2001), 195–211.
77. P.-L. Lions, N. Masmoudi: *Une approche locale de la limite incompressible*, C. R. Acad. Sci. Paris Sr. I Math. **329** (1999), 387–392.
78. T. P. Liu. *Solutions in the large for the equations of nonisentropic gas dynamics*, Indiana Univ. Math. J. **26** (1977), 147–177.
79. T.-P. Liu, S.-H. Yu: *Boltzmann equation: micro-macro decompositions and positivity of shock profiles*, Comm. Math. Phys. **246** (2004), 133–179.
80. T.-P. Liu, T. Yang, S.-H. Yu: *Energy method for Boltzmann equation*, Phys. D **188** (2004), 178–192.
81. E. Mach. *Die Mechanik in ihrer Entwickelung.* Leipzig, zweite Auflage, 1889.
82. N. Masmoudi, L. Saint-Raymond. From the Boltzmann equation to the Stokes-Fourier system in a bounded domain, *Comm. Pure Appl. Math.*, **56** (2003), 1263–1293.
83. G. Métivier, K. Zumbrun. *Existence of semilinear relaxation shocks.* Preprint 2008.

84. S. Mischler. *On the initial boundary value problem for the Vlasov-Poisson-Boltzmann system*, Comm. Math. Phys. **210** (2000), 447–466.
85. S. Mischler. *Kinetic equations with Maxwell boundary condition*, Preprint (2002).
86. C. B. Morrey. *On the derivation of the equations of hydrodynamics from Statistical Mechanics*, Commun. Pure Appl. Math., **8** (1955), 279–290.
87. C. Mouhot. *Rate of convergence to equilibrium for the spatially homogeneous Boltzmann equation with hard potentials,* Comm. Math. Phys. **261** (2006), 629–672.
88. T. Nishida. *Fluid dynamical limit of the nonlinear Boltzmann equation to the level of the compressible Euler equation.* Comm. Math. Phys. **61** (1978), 119–148.
89. S. Olla, S. Varadhan, H. Yau, *Hydrodynamical limit for a Hamiltonian system with weak noise*, Commun. Math. Phys. **155** (1993), 523–560.
90. B. Perthame: *Introduction to the collision models in Boltzmann's theory*, in "Modeling of Collisions", P.-A. Raviart ed., Masson, Paris, 1997.
91. J. Quastel and H.-T. Yau, *Lattice gases, large deviations, and the incompressible Navier-Stokes equations*, Ann. of Math. **148** (1998), 51–108.
92. L. Saint-Raymond: *From the BGK model to the Navier-Stokes equations*, Ann. Sci. Ecole Norm. Sup. (4) **36** (2003), 271–317.
93. L. Saint-Raymond: *Du modèle BGK de l'équation de Boltzmann aux équations d'Euler des fluides incompressibles*, Bull. Sci. Math. **126** (2002), 493–506.
94. L. Saint-Raymond: *Convergence of solutions to the Boltzmann equation in the incompressible Euler limit*, Arch. Ration. Mech. Anal. **166** (2003), 47–80.
95. L. Saint-Raymond: *Hydrodynamic limits: some improvements of the relative entropy method*, Ann. Inst. H. Poincaré, to appear, 2008. doi:10.1016/j.anihpc.2008.01.001.
96. S. Schochet. *Fast singular limits of hyperbolic PDEs,* J. Differential Equations **114** (1994), 476–512.
97. D. Serre: "Systèmes de lois de conservation. I. Hyperbolicité, entropies, ondes de choc. Fondations." Diderot Editeur, Paris, 1996.
98. D. Serre: "Systèmes de lois de conservation. II. Structures géométriques, oscillation et problèmes mixtes. Fondations." Diderot Editeur, Paris, 1996.
99. J. Serrin: *On the interior regularity of weak solutions of the Navier-Stokes equations.* Arch. Rational Mech. Anal. **9** (1962), 187–195.
100. T. Sideris: *Formation of Singularities in 3D Compressible Fluids*, Commun. Math. Phys. **101** (1985), 475–485.
101. J. Smoller: "Shock waves and reaction-diffusion equations." Grundlehren der Mathematischen Wissenschaften, Springer-Verlag, New York-Berlin, 1983.
102. R. Temam: "Navier-Stokes equations. Theory and numerical analysis." Studies in Mathematics and its Applications, North-Holland Publishing Co., Amsterdam-New York-Oxford, 1977.
103. S. Ukai. *On the existence of global solutions of a mixed problem for the nonlinear Boltzmann equation*, Proc. of the Japan Acad. **50** (1974), 179–184.
104. S. Ukai: *Eigenvalues of the neutron transport operator for a homogeneous finite moderator*, J. Math. Anal. Appl. **18** (1967), 297–314.
105. S. Ukai, K. Asano: *The Euler limit and initial layer of the nonlinear Boltzmann equation*, Hokkaido Math. J. **12** (1983), 311–332.
106. C. Villani: "A review of mathematical topics in collisional kinetic theory." Handbook of mathematical fluid dynamics, North-Holland, Amsterdam, 2002.
107. P. Volkmann. *Einführung in das Studium der theoretischen Physik.* Leipzig, 1900.

108. H. T. Yau. *Relative entropy and hydrodynamics of Ginzburg-Landau models*, Lett. Math. Phys. **22** (1991), 63–80.
109. H. T. Yau. *Scaling limit of particle systems, incompressible Navier-Stokes equation and Boltzmann equation.* Proceedings of the International Congress of Mathematicians, Doc. Math. **3** (1998), 193–202.
110. V. Yudovitch. *Non stationnary flows of an ideal incompressible fluid*, Zh. Vych. Math. **3** (1963), 1032–1066.

Index

Lecture Notes in Mathematics

For information about earlier volumes
please contact your bookseller or Springer
LNM Online archive: springerlink.com

Vol. 1830: M. I. Gil', Operator Functions and Localization of Spectra. XIV, 256 p, 2003.

Vol. 1831: A. Connes, J. Cuntz, E. Guentner, N. Higson, J. E. Kaminker, Noncommutative Geometry, Martina Franca, Italy 2002. Editors: S. Doplicher, L. Longo (2004)

Vol. 1832: J. Azéma, M. Émery, M. Ledoux, M. Yor (Eds.), Séminaire de Probabilités XXXVII (2003)

Vol. 1833: D.-Q. Jiang, M. Qian, M.-P. Qian, Mathematical Theory of Nonequilibrium Steady States. On the Frontier of Probability and Dynamical Systems. IX, 280 p, 2004.

Vol. 1834: Yo. Yomdin, G. Comte, Tame Geometry with Application in Smooth Analysis. VIII, 186 p, 2004.

Vol. 1835: O.T. Izhboldin, B. Kahn, N.A. Karpenko, A. Vishik, Geometric Methods in the Algebraic Theory of Quadratic Forms. Summer School, Lens, 2000. Editor: J.-P. Tignol (2004)

Vol. 1836: C. Năstăsescu, F. Van Oystaeyen, Methods of Graded Rings. XIII, 304 p, 2004.

Vol. 1837: S. Tavaré, O. Zeitouni, Lectures on Probability Theory and Statistics. Ecole d'Eté de Probabilités de Saint-Flour XXXI-2001. Editor: J. Picard (2004)

Vol. 1838: A.J. Ganesh, N.W. O'Connell, D.J. Wischik, Big Queues. XII, 254 p, 2004.

Vol. 1839: R. Gohm, Noncommutative Stationary Processes. VIII, 170 p, 2004.

Vol. 1840: B. Tsirelson, W. Werner, Lectures on Probability Theory and Statistics. Ecole d'Eté de Probabilités de Saint-Flour XXXII-2002. Editor: J. Picard (2004)

Vol. 1841: W. Reichel, Uniqueness Theorems for Variational Problems by the Method of Transformation Groups (2004)

Vol. 1842: T. Johnsen, A. L. Knutsen, K_3 Projective Models in Scrolls (2004)

Vol. 1843: B. Jefferies, Spectral Properties of Noncommuting Operators (2004)

Vol. 1844: K.F. Siburg, The Principle of Least Action in Geometry and Dynamics (2004)

Vol. 1845: Min Ho Lee, Mixed Automorphic Forms, Torus Bundles, and Jacobi Forms (2004)

Vol. 1846: H. Ammari, H. Kang, Reconstruction of Small Inhomogeneities from Boundary Measurements (2004)

Vol. 1847: T.R. Bielecki, T. Björk, M. Jeanblanc, M. Rutkowski, J.A. Scheinkman, W. Xiong, Paris-Princeton Lectures on Mathematical Finance 2003 (2004)

Vol. 1848: M. Abate, J. E. Fornaess, X. Huang, J. P. Rosay, A. Tumanov, Real Methods in Complex and CR Geometry, Martina Franca, Italy 2002. Editors: D. Zaitsev, G. Zampieri (2004)

Vol. 1849: Martin L. Brown, Heegner Modules and Elliptic Curves (2004)

Vol. 1850: V. D. Milman, G. Schechtman (Eds.), Geometric Aspects of Functional Analysis. Israel Seminar 2002-2003 (2004)

Vol. 1851: O. Catoni, Statistical Learning Theory and Stochastic Optimization (2004)

Vol. 1852: A.S. Kechris, B.D. Miller, Topics in Orbit Equivalence (2004)

Vol. 1853: Ch. Favre, M. Jonsson, The Valuative Tree (2004)

Vol. 1854: O. Saeki, Topology of Singular Fibers of Differential Maps (2004)

Vol. 1855: G. Da Prato, P.C. Kunstmann, I. Lasiecka, A. Lunardi, R. Schnaubelt, L. Weis, Functional Analytic Methods for Evolution Equations. Editors: M. Iannelli, R. Nagel, S. Piazzera (2004)

Vol. 1856: K. Back, T.R. Bielecki, C. Hipp, S. Peng, W. Schachermayer, Stochastic Methods in Finance, Bressanone/Brixen, Italy, 2003. Editors: M. Fritelli, W. Runggaldier (2004)

Vol. 1857: M. Émery, M. Ledoux, M. Yor (Eds.), Séminaire de Probabilités XXXVIII (2005)

Vol. 1858: A.S. Cherny, H.-J. Engelbert, Singular Stochastic Differential Equations (2005)

Vol. 1859: E. Letellier, Fourier Transforms of Invariant Functions on Finite Reductive Lie Algebras (2005)

Vol. 1860: A. Borisyuk, G.B. Ermentrout, A. Friedman, D. Terman, Tutorials in Mathematical Biosciences I. Mathematical Neurosciences (2005)

Vol. 1861: G. Benettin, J. Henrard, S. Kuksin, Hamiltonian Dynamics – Theory and Applications, Cetraro, Italy, 1999. Editor: A. Giorgilli (2005)

Vol. 1862: B. Helffer, F. Nier, Hypoelliptic Estimates and Spectral Theory for Fokker-Planck Operators and Witten Laplacians (2005)

Vol. 1863: H. Führ, Abstract Harmonic Analysis of Continuous Wavelet Transforms (2005)

Vol. 1864: K. Efstathiou, Metamorphoses of Hamiltonian Systems with Symmetries (2005)

Vol. 1865: D. Applebaum, B.V. R. Bhat, J. Kustermans, J. M. Lindsay, Quantum Independent Increment Processes I. From Classical Probability to Quantum Stochastic Calculus. Editors: M. Schürmann, U. Franz (2005)

Vol. 1866: O.E. Barndorff-Nielsen, U. Franz, R. Gohm, B. Kümmerer, S. Thorbjønsen, Quantum Independent Increment Processes II. Structure of Quantum Lévy Processes, Classical Probability, and Physics. Editors: M. Schürmann, U. Franz, (2005)

Vol. 1867: J. Sneyd (Ed.), Tutorials in Mathematical Biosciences II. Mathematical Modeling of Calcium Dynamics and Signal Transduction. (2005)

Vol. 1868: J. Jorgenson, S. Lang, $Pos_n(R)$ and Eisenstein Series. (2005)

Vol. 1869: A. Dembo, T. Funaki, Lectures on Probability Theory and Statistics. Ecole d'Eté de Probabilités de Saint-Flour XXXIII-2003. Editor: J. Picard (2005)

Vol. 1870: V.I. Gurariy, W. Lusky, Geometry of Müntz Spaces and Related Questions. (2005)

Vol. 1871: P. Constantin, G. Gallavotti, A.V. Kazhikhov, Y. Meyer, S. Ukai, Mathematical Foundation of Turbulent Viscous Flows, Martina Franca, Italy, 2003. Editors: M. Cannone, T. Miyakawa (2006)

Vol. 1872: A. Friedman (Ed.), Tutorials in Mathematical Biosciences III. Cell Cycle, Proliferation, and Cancer (2006)

Vol. 1873: R. Mansuy, M. Yor, Random Times and Enlargements of Filtrations in a Brownian Setting (2006)

Vol. 1874: M. Yor, M. Émery (Eds.), In Memoriam Paul-André Meyer - Séminaire de Probabilités XXXIX (2006)

Vol. 1875: J. Pitman, Combinatorial Stochastic Processes. Ecole d'Eté de Probabilités de Saint-Flour XXXII-2002. Editor: J. Picard (2006)

Vol. 1876: H. Herrlich, Axiom of Choice (2006)

Vol. 1877: J. Steuding, Value Distributions of L-Functions (2007)

Vol. 1878: R. Cerf, The Wulff Crystal in Ising and Percolation Models, Ecole d'Eté de Probabilités de Saint-Flour XXXIV-2004. Editor: Jean Picard (2006)

Vol. 1879: G. Slade, The Lace Expansion and its Applications, Ecole d'Eté de Probabilités de Saint-Flour XXXIV-2004. Editor: Jean Picard (2006)

Vol. 1880: S. Attal, A. Joye, C.-A. Pillet, Open Quantum Systems I, The Hamiltonian Approach (2006)

Control Theory. Cetraro, Italy 2004. Editors: P. Nistri, G. Stefani (2008)

Vol. 1933: M. Petkovic, Point Estimation of Root Finding Methods (2008)

Vol. 1934: C. Donati-Martin, M. Émery, A. Rouault, C. Stricker (Eds.), Séminaire de Probabilités XLI (2008)

Vol. 1935: A. Unterberger, Alternative Pseudodifferential Analysis (2008)

Vol. 1936: P. Magal, S. Ruan (Eds.), Structured Population Models in Biology and Epidemiology (2008)

Vol. 1937: G. Capriz, P. Giovine, P.M. Mariano (Eds.), Mathematical Models of Granular Matter (2008)

Vol. 1938: D. Auroux, F. Catanese, M. Manetti, P. Seidel, B. Siebert, I. Smith, G. Tian, Symplectic 4-Manifolds and Algebraic Surfaces. Cetraro, Italy 2003. Editors: F. Catanese, G. Tian (2008)

Vol. 1939: D. Boffi, F. Brezzi, L. Demkowicz, R.G. Durán, R.S. Falk, M. Fortin, Mixed Finite Elements, Compatibility Conditions, and Applications. Cetraro, Italy 2006. Editors: D. Boffi, L. Gastaldi (2008)

Vol. 1940: J. Banasiak, V. Capasso, M.A.J. Chaplain, M. Lachowicz, J. Miękisz, Multiscale Problems in the Life Sciences. From Microscopic to Macroscopic. Będlewo, Poland 2006. Editors: V. Capasso, M. Lachowicz (2008)

Vol. 1941: S.M.J. Haran, Arithmetical Investigations. Representation Theory, Orthogonal Polynomials, and Quantum Interpolations (2008)

Vol. 1942: S. Albeverio, F. Flandoli, Y.G. Sinai, SPDE in Hydrodynamic. Recent Progress and Prospects. Cetraro, Italy 2005. Editors: G. Da Prato, M. Röckner (2008)

Vol. 1943: L.L. Bonilla (Ed.), Inverse Problems and Imaging. Martina Franca, Italy 2002 (2008)

Vol. 1944: A. Di Bartolo, G. Falcone, P. Plaumann, K. Strambach, Algebraic Groups and Lie Groups with Few Factors (2008)

Vol. 1945: F. Brauer, P. van den Driessche, J. Wu (Eds.), Mathematical Epidemiology (2008)

Vol. 1946: G. Allaire, A. Arnold, P. Degond, T.Y. Hou, Quantum Transport. Modelling, Analysis and Asymptotics. Cetraro, Italy 2006. Editors: N.B. Abdallah, G. Frosali (2008)

Vol. 1947: D. Abramovich, M. Mariño, M. Thaddeus, R. Vakil, Enumerative Invariants in Algebraic Geometry and String Theory. Cetraro, Italy 2005. Editors: K. Behrend, M. Manetti (2008)

Vol. 1948: F. Cao, J-L. Lisani, J-M. Morel, P. Musé, F. Sur, A Theory of Shape Identification (2008)

Vol. 1949: H.G. Feichtinger, B. Helffer, M.P. Lamoureux, N. Lerner, J. Toft, Pseudo-Differential Operators. Quantization and Signals. Cetraro, Italy 2006. Editors: L. Rodino, M.W. Wong (2008)

Vol. 1950: M. Bramson, Stability of Queueing Networks, Ecole d'Eté de Probabilités de Saint-Flour XXXVI-2006 (2008)

Vol. 1951: A. Moltó, J. Orihuela, S. Troyanski, M. Valdivia, A Non Linear Transfer Technique for Renorming (2009)

Vol. 1952: R. Mikhailov, I.B.S. Passi, Lower Central and Dimension Series of Groups (2009)

Vol. 1953: K. Arwini, C.T.J. Dodson, Information Geometry (2008)

Vol. 1954: P. Biane, L. Bouten, F. Cipriani, N. Konno, N. Privault, Q. Xu, Quantum Potential Theory. Editors: U. Franz, M. Schuermann (2008)

Vol. 1955: M. Bernot, V. Caselles, J.-M. Morel, Optimal Transportation Networks (2008)

Vol. 1956: C.H. Chu, Matrix Convolution Operators on Groups (2008)

Vol. 1957: A. Guionnet, On Random Matrices: Macroscopic Asymptotics, Ecole d'Eté de Probabilités de Saint-Flour XXXVI-2006 (2009)

Vol. 1958: M.C. Olsson, Compactifying Moduli Spaces for Abelian Varieties (2008)

Vol. 1959: Y. Nakkajima, A. Shiho, Weight Filtrations on Log Crystalline Cohomologies of Families of Open Smooth Varieties (2008)

Vol. 1960: J. Lipman, M. Hashimoto, Foundations of Grothendieck Duality for Diagrams of Schemes (2009)

Vol. 1961: G. Buttazzo, A. Pratelli, S. Solimini, E. Stepanov, Optimal Urban Networks via Mass Transportation (2009)

Vol. 1962: R. Dalang, D. Khoshnevisan, C. Mueller, D. Nualart, Y. Xiao, A Minicourse on Stochastic Partial Differential Equations (2009)

Vol. 1963: W. Siegert, Local Lyapunov Exponents (2009)

Vol. 1964: W. Roth, Operator-valued Measures and Integrals for Cone-valued Functions and Integrals for Cone-valued Functions (2009)

Vol. 1965: C. Chidume, Geometric Properties of Banach Spaces and Nonlinear Iterations (2009)

Vol. 1966: D. Deng, Y. Han, Harmonic Analysis on Spaces of Homogeneous Type (2009)

Vol. 1967: B. Fresse, Modules over Operads and Functors (2009)

Vol. 1968: R. Weissauer, Endoscopy for GSP(4) and the Cohomology of Siegel Modular Threefolds (2009)

Vol. 1969: B. Roynette, M. Yor, Penalising Brownian Paths (2009)

Vol. 1970: R. Kotecký, Methods of Contemporary Mathematical Statistical Physics (2009)

Vol. 1971: L. Saint-Raymond, Hydrodynamic Limits of the Boltzmann Equation (2009)

Vol. 1972: T. Mochizuki, Donaldson Type Invariants for Algebraic Surfaces (2009)

Recent Reprints and New Editions

Vol. 1702: J. Ma, J. Yong, Forward-Backward Stochastic Differential Equations and their Applications. 1999 – Corr. 3rd printing (2007)

Vol. 830: J.A. Green, Polynomial Representations of GL_n, with an Appendix on Schensted Correspondence and Littelmann Paths by K. Erdmann, J.A. Green and M. Schoker 1980 – 2nd corr. and augmented edition (2007)

Vol. 1693: S. Simons, From Hahn-Banach to Monotonicity (Minimax and Monotonicity 1998) – 2nd exp. edition (2008)

Vol. 470: R.E. Bowen, Equilibrium States and the Ergodic Theory of Anosov Diffeomorphisms. With a preface by D. Ruelle. Edited by J.-R. Chazottes. 1975 – 2nd rev. edition (2008)

Vol. 523: S.A. Albeverio, R.J. Høegh-Krohn, S. Mazzucchi, Mathematical Theory of Feynman Path Integral. 1976 – 2nd corr. and enlarged edition (2008)

Vol. 1764: A. Cannas da Silva, Lectures on Symplectic Geometry 2001 – Corr. 2nd printing (2008)

LECTURE NOTES IN MATHEMATICS  Springer

Edited by J.-M. Morel, F. Takens, B. Teissier, P.K. Maini

Editorial Policy (for the publication of monographs)

1. Lecture Notes aim to report new developments in all areas of mathematics and their applications - quickly, informally and at a high level. Mathematical texts analysing new developments in modelling and numerical simulation are welcome.

   Monograph manuscripts should be reasonably self-contained and rounded off. Thus they may, and often will, present not only results of the author but also related work by other people. They may be based on specialised lecture courses. Furthermore, the manuscripts should provide sufficient motivation, examples and applications. This clearly distinguishes Lecture Notes from journal articles or technical reports which normally are very concise. Articles intended for a journal but too long to be accepted by most journals, usually do not have this "lecture notes" character. For similar reasons it is unusual for doctoral theses to be accepted for the Lecture Notes series, though habilitation theses may be appropriate.

2. Manuscripts should be submitted either online at www.editorialmanager.com/lnm to Springer's mathematics editorial in Heidelberg, or to one of the series editors. In general, manuscripts will be sent out to 2 external referees for evaluation. If a decision cannot yet be reached on the basis of the first 2 reports, further referees may be contacted: The author will be informed of this. A final decision to publish can be made only on the basis of the complete manuscript, however a refereeing process leading to a preliminary decision can be based on a pre-final or incomplete manuscript. The strict minimum amount of material that will be considered should include a detailed outline describing the planned contents of each chapter, a bibliography and several sample chapters.

   Authors should be aware that incomplete or insufficiently close to final manuscripts almost always result in longer refereeing times and nevertheless unclear referees' recommendations, making further refereeing of a final draft necessary.

   Authors should also be aware that parallel submission of their manuscript to another publisher while under consideration for LNM will in general lead to immediate rejection.

3. Manuscripts should in general be submitted in English. Final manuscripts should contain at least 100 pages of mathematical text and should always include

   - a table of contents;
   - an informative introduction, with adequate motivation and perhaps some historical remarks: it should be accessible to a reader not intimately familiar with the topic treated;
   - a subject index: as a rule this is genuinely helpful for the reader.

   For evaluation purposes, manuscripts may be submitted in print or electronic form (print form is still preferred by most referees), in the latter case preferably as pdf- or zipped ps-files. Lecture Notes volumes are, as a rule, printed digitally from the authors' files. To ensure best results, authors are asked to use the LaTeX2e style files available from Springer's web-server at:

   ftp://ftp.springer.de/pub/tex/latex/svmonot1/ (for monographs) and
   ftp://ftp.springer.de/pub/tex/latex/svmultt1/ (for summer schools/tutorials).

Additional technical instructions, if necessary, are available on request from: lnm@springer.com.

4. Careful preparation of the manuscripts will help keep production time short besides ensuring satisfactory appearance of the finished book in print and online. After acceptance of the manuscript authors will be asked to prepare the final LaTeX source files and also the corresponding dvi-, pdf- or zipped ps-file. The LaTeX source files are essential for producing the full-text online version of the book (see http://www.springerlink.com/openurl.asp?genre=journal&issn=0075-8434 for the existing online volumes of LNM).

    The actual production of a Lecture Notes volume takes approximately 12 weeks.

5. Authors receive a total of 50 free copies of their volume, but no royalties. They are entitled to a discount of 33.3% on the price of Springer books purchased for their personal use, if ordering directly from Springer.

6. Commitment to publish is made by letter of intent rather than by signing a formal contract. Springer-Verlag secures the copyright for each volume. Authors are free to reuse material contained in their LNM volumes in later publications: a brief written (or e-mail) request for formal permission is sufficient.

Addresses:

Professor J.-M. Morel, CMLA,
École Normale Supérieure de Cachan,
61 Avenue du Président Wilson, 94235 Cachan Cedex, France
E-mail: Jean-Michel.Morel@cmla.ens-cachan.fr

Professor F. Takens, Mathematisch Instituut,
Rijksuniversiteit Groningen, Postbus 800,
9700 AV Groningen, The Netherlands
E-mail: F.Takens@rug.nl

Professor B. Teissier, Institut Mathématique de Jussieu,
UMR 7586 du CNRS, Équipe "Géométrie et Dynamique",
175 rue du Chevaleret,
75013 Paris, France
E-mail: teissier@math.jussieu.fr

For the "Mathematical Biosciences Subseries" of LNM:

Professor P.K. Maini, Center for Mathematical Biology,
Mathematical Institute, 24-29 St Giles,
Oxford OX1 3LP, UK
E-mail: maini@maths.ox.ac.uk

Springer, Mathematics Editorial, Tiergartenstr. 17,
69121 Heidelberg, Germany,
Tel.: +49 (6221) 487-259
Fax: +49 (6221) 4876-8259
E-mail: lnm@springer.com